# Efficiency and Power in Energy Conversion and Storage

## Basic Physical Concepts

# Efficiency and Power in Energy Conversion and Storage

## Basic Physical Concepts

Thomas Christen

CRC Press
Taylor & Francis Group
Boca Raton  London  New York

CRC Press is an imprint of the
Taylor & Francis Group, an **informa** business

CRC Press
Taylor & Francis Group
6000 Broken Sound Parkway NW, Suite 300
Boca Raton, FL 33487-2742

First issued in paperback 2023

**Library of Congress Cataloging-in-Publication Data**

Names: Christen, Thomas, 1963- author.
Title: Efficiency and power in energy conversion and storage : basic physical concepts / Thomas Christen.
Description: First edition. | Boca Raton, FL : CRC Press/Taylor & Francis Group, 2018. | "A CRC title, part of the Taylor & Francis imprint, a member of the Taylor & Francis Group, the academic division of T&F Informa plc." | Includes bibliographical references and index.
Identifiers: LCCN 2018015849 (print) | LCCN 2018016864 (ebook) | ISBN 9780429454288 (eBook) | ISBN 9780429845253 (Adobe PDF) | ISBN 9780429845246 (ePUB) | ISBN 9780429845239 (Mobipocket) | ISBN 9781138626638 (hardback : acid-free paper)
Subjects: LCSH: Power resources. | Energy conversion--Mathematics. | Energy storage--Mathematics. | Electric power plants--Efficiency.
Classification: LCC TJ163.2 (ebook) | LCC TJ163.2 .C8827 2018 (print) | DDC 621.042--dc23
LC record available at https://lccn.loc.gov/2018015849

**Visit the Taylor & Francis Web site at
http://www.taylorandfrancis.com**

**and the CRC Press Web site at
http://www.crcpress.com**

ISBN 13: 978-1-032-65250-4 (pbk)
ISBN 13: 978-1-138-62663-8 (hbk)
ISBN 13: 978-0-429-45428-8 (ebk)

DOI: 10.1201/9780429454288

# Contents

# Foreword

This book aims to bridge the gap between scientific basics and engineering applications in energy conversion and storage in a concise manner. It evolved from lectures for students with a basic knowledge of the educational canon of physics on a typical level for a bachelor's degree in science and engineering.

It is needless to re-emphasize the actual relevance and importance of energy issues arising due to the huge ecological and economical challenges associated with mankind's growing energy demand and consumption, due to recent and ongoing progress in different energy conversion and storage technologies, and due to the change in electrical power infrastructure. Knowledge of the underlying physics and modeling know-how are prerequisites for optimizing the ecological and economical performance of related technologies and products within industrial and academic research and development. I tried to keep the book short while covering the relevant concepts. The *efficiency-power* or *energy-power relations* are among the most important characteristics of energy conversion and storage systems. They are used for quantifying the trade-off between high power and high efficiency, they serve as the recurrent theme in this book and will be illustrated with practically relevant, but simple and analytically accessible, examples.

After the Introduction, the chapter on Equilibrium Statistical Thermodynamics reviews the theory of equilibrium states and processes, which builds the basis for the later chapters. The remainder, nonequilibrium issues and various application examples, is thought of as a primer. Almost all equations are derived more or less from scratch, in a way such that not much paper and pencil will be wasted by those readers who want to reproduce the derivations. Although mathematical symbols for physical quantities are often defined according to common usage in the literature and are thus not unique (e.g., $p$ for pressure, momentum, normalized power; $f$ for distribution function, molecular degrees of freedom, dummy functions, etc.,) their meanings should always be clear from the context without creating confusion. The examples might seem oversimplified and the figures rather simplistic, because of the attempt to keep them devoid of details that are believed to be irrelevant for understanding the bare essentials. But once the simple cases are understood in depth, it will be straightforward for the reader to apply the concepts to realistic and more complicated cases.

I am very grateful to Christian Ohler and Martin Carlen for inspiring discussions and collaboration, and to Frank Kassubek, Jaroslav Hemrle, Stephan Schnez, and Uwe Drofenik for many valuable remarks and comments on various chapters. Although most of the discussed topics belong to common knowledge and originality of the presented results is not at all claimed, I take the full responsibility for possible criticism.

Thomas Christen
Address:
ABB Switzerland Ltd, Corporate Research
Segelhofstrasse 1K, CH-5405 Baden-Dättwil
Switzerland

# 1 Introduction

This chapter briefly introduces the common theme of the book: efficiency-power relations. The technical terms and notions used here will be defined in later chapters in more detail.

## 1.1 ENERGY, POWER, AND EFFICIENCY

An essential insight of contemporary science and engineering is the *conservation of energy* and the occurrence of energy in various different forms, like heat, electricity, magnetism, elasticity, gravitation, kinetic energy etc., up to Einstein's famous relativistic energy-mass equivalence. Energy as a property refers to the ability to deliver work. Energy as a quantity can neither be created nor destroyed. It can be either converted between different forms, stored in specific forms, or transported. Energy *consumption* is thus nothing else than a *conversion* or *transformation* of one form of energy into another one, be it work or heat. You might say, it is a good thing that energy cannot be destroyed. However, in many cases the final energy form after a conversion process has less value for usage than the initial form, namely, when heat, or entropy, is generated, as will be discussed later. So the bad message is that energy available for work (sometimes called *exergy*) *can* be destroyed. The obviously important ratio of energy which remains for use to the total energy is related to the *efficiency*.

It is not only the amount and the form of energy that is relevant for energy consumption: the consumer needs the energy within a certain *time*. The amount of energy converted per time, the *power*, is thus often the main quantity of a request. We all know from our daily experience that it can be difficult to harvest a high power from a given amount of energy without reducing the efficiency. The trade-off problem associated with these two conflicting objectives, high power and high efficiency, forms the common theme of this book. The *efficiency-power relations* turn out to play a central role in characterizing all kinds of physical, chemical, biological, and technical energy conversion and storage systems. The determination of the efficiency-power relation requires an understanding of fundamental concepts like reversible and irreversible processes, losses in conversion and transport processes, etc.. In this book we shall focus on this understanding by unveiling the basic theory and providing some simple application examples for illustration. For an exhaustive discussion including many realistic applications in detail, you should take the time for studying more extensive monographs like Refs. [Bej88, Hug10, Str14, Ruf17], to mention a few.

Energy conservation and the decrease of its useability imply that energy systems are naturally described by an *energy flow cascade*. Two examples

| Largest cosmological /astronomical events | $10^{50}$ W |
|---|---|
| Power flow from sun to earth | $10^{17}$ W |
| Mankind's energy consumption | $10^{13}$ W |
| Largest power plants (hydro) | $10^{10}$ W |
| Largest power devices (motors, generators, turbines, ...) | $10^{9}$ W |
| Wind turbine (typical, horizontal axis) | $10^{6}$ W |
| Vehicle engine, flywheel | $10^{4}$-$10^{5}$ W |
| Average consumption of a human being | $10^{3}$ W |
| Household equipment (cooking plate, microwave, hair dryer, ...) | $10^{2}$-$10^{3}$ W |
| Total metabolic power of a human being, laptop computer | $10^{2}$ W |
| Standby power of appliances | 1-10 W |
| Human heart (average, 1 J at one beat per second) | 1 W |
| Wireless router | $10^{-2}$ W |
| Typical phone transmission power (UMTS) | $10^{-3}$ W |
| Wristwatch | $10^{-6}$ W |
| Cell metabolism | $10^{-12}$ W |
| Molecular motors | $10^{-15}$ W |

**Table 1.1**
**Power values for various examples.**

which immediately come into mind are the biosphere, where the energy flow cascade is required for the preservation and evolution of life, and mankind's energy infrastructure for heating, electricity, industrial processes, mobility, etc.. Many interesting things could be said about the former, from a global ecosystem perspective down to cell metabolism. A special property of the biosphere is its *far from thermodynamic equilibrium state*, involving cycles of matter like carbon, nitrogen, oxygen, and water cycles, which makes an understanding rather difficult. On the other hand, technical devices in our energy and power infrastructure often run in states that deviate only weakly from thermodynamic equilibrium, and can thus be understood rather easily. This book is restricted mainly to *technical devices*. But one should be aware that the biosphere and technology are strongly coupled and overlapping, for instance in the context of technological contributions to $CO_2$-emission and global warming, and its use for agriculture and food processing, etc.; in fact we and our technologies are just a part of the biosphere. Last but not least, one can often *learn* from biology, e.g., for designing technical devices.

The energy flow cascade associated with our energy consumption is reflected in the hierarchy of energy forms according to their position in the conversion chain of our energy infrastructure: *primary energy, secondary energy,* and *end use energy*. This classification is partly anthropocentric, because primary energy is defined as input into our conversion chain, i.e., as not being converted yet by a human technology. Nuclear energies from fission and fusion, geothermal heat, solar energy and its consequences of wind and water flow (hydro-power; can of course also be related to the tides related to the gravitation and kinetic energies of the earth-moon system), and the chemical energy of crude oil, gas and coal - they are all primary energies. *Secondary energy* forms are those which result from primary sources by transformation with the help of energy conversion devices. The two most important secondary energies are certainly heat and electricity. *End use energy* refers to applications like heating, cooling, lighting, motion and transportation, vacuum cleaning, ... used by the consumer for industrial, public, or private use. At the final end of the chain only heat remains. Energy conversion, storage, and transport occur on all levels, and every device (or system or subsystem) can be characterized by its efficiency, power capability and energy storage capacity.

In order to get a feel for orders of magnitudes of energy, power, and efficiency values, it can be helpful to consult the actual information sources on energy consumption statistics and the literature on specific technologies. Numbers connected to our energy usage and supply technologies are continuously changing over time and updated. Many details can be found in the excellent books [Mac09, Smi17, Ruf17]. Tables 1.1 and 1.2 list some rough power and efficiency values which give an impression of typical orders of magnitudes. Due to the extensive character of the energy, for applications it is often useful to consider densities or specific quantities, e.g., with units $J/m^3$ or $J/kg$, for characterizing specific energy conversion and storage devices, particularly for comparison and scaling issues. Most convenient is the unit $MJ/kg$ for chemical energies (see Table 1.3). In regards to power, not only density $W/m^3$ or the specific value $W/kg$ are of interest for the purpose of scaling issues, but also the *energy current density* in units of $W/m^2$, which characterizes energy transport, plays an important role (like the terrestrial solar energy current with order of magnitude $1 \ kW/m^2$).

Because the *efficiency* is the ratio of the energy *of use* to the energy input, its value depends somewhat on the context which has thus to be clearly specified. You should always be suspicious when efficiencies for different systems are compared to each other without specifying the context. The most famous efficiency is certainly the *Carnot efficiency* $\eta = 1 - T_2/T_1$ for heat engines which run in thermodynamic equilibrium (i.e., reversibly) between hot and cold heat baths at temperatures $T_1$ and $T_2$, respectively. A further reduction of the useful energy can be due to irreversibilities like friction or leakage losses. There exist thus two different types of efficiency reductions: the *fundamental*, unavoidable ones which occur even for reversible processes, and those related to nonequilibrium processes which may be partly mitigated, at least with some effort. You certainly have learned earlier that reversible processes run, in theory, infinitely slowly, which means at zero power. As soon as you take a finite power, irreversible losses appear, and they usually increase with power.

| Largest transformers, motors, generators (electrical) | 99% |
|---|---|
| Large hydro power plants (electrical) | 90-95% |
| Fuel cells (electrical) | 40-50% |
| Wind turbine (electrical) | 30-40% |
| Nuclear and thermal power plants (electrical) | 30-40% |
| World electricity generation (electrical) | 40% |
| Gasoline/diesel car combustion engine (mechanical) | 20-30% |
| Refrigerator (cooling) | 20-50% |
| Solar thermal (electrical) | 20-30% |
| Microturbine ($< 100$ kW) (electrical) | 20% |
| Bio-efficiency, metabolism, swimming, running, flying | 15-50% |
| Photovoltaic cells (electrical) | 10-20% |
| Photosynthesis (chemical) | $< 5\%$ |
| Incandescent light bulbs (luminous) | 2% |

**Table 1.2**
**Typical efficiency values for various examples.**

All this and more turns out to be related to the efficiency-power relations, and will be discussed in later chapters in detail. The following section provides a simple forward-looking example.

## 1.2   THE EFFICIENCY-POWER TRADE-OFF

The main questions associated with energy usage may be formulated as follows: *"How much of a given amount of energy is available for a specific use?"* and *"At which rate can one get it?"* These questions are connected to the partly contradicting tasks of *maximizing the efficiency $\eta$* and *the power $P$* at the same time.

Consider the simple electric circuit in Figure 1.1(a). A consumer resistance $R$ draws a power $P = \mathsf{U}I$ from a battery with constant open-circuit voltage $\mathsf{U}_0$ and internal resistance $R_0$. (Further on, $\mathsf{U}$ is used for voltage, since $U$ will be reserved for inner energy in thermodynamics.) We neglect, for the moment, internal leakage of the battery (dashed leakage resistor in the illustration). Kirchhoff's second rule together with Ohm's law, $\mathsf{U} = RI$, tells us that the current $I$ obeys $\mathsf{U}_0 = (R + R_0)I$. Because the total power provided by the battery is $P_{tot} = \mathsf{U}_0 I$, the efficiency, defined by

$$\eta = \frac{P}{P_{tot}} \ , \tag{1.1}$$

| Relativistic energy of mass ($c^2$) | $10^{17}$ J/kg |
|---|---|
| Nuclear power fuel | $10^{12}$ J/kg |
| Energy intensities of materials (energetic costs of fabrication) | $10^7$-$10^9$ J/kg |
| Chemical energy of fuels (gas, coal, oil, wood) and nutrition | $10^7$-$10^8$ J/kg |
| Latent heat | $10^5$-$10^6$ J/kg |
| Sensible heat of matter per Kelvin | $10^3$ J/kg/K |
| Gravitational energy per height (on earth) ($g$) | 10 J/kg/m |
| Electric motors and generators | $10^3$-$10^4$ W/kg |
| Heat production in human body | 1-2 W/kg |
| Solar heat current density | $10^3$ W/m$^2$ |
| Typical heat current density from geothermal fields | 10 W/m$^2$ |
| Various energy storage devices | see Figure 6.1 |

**Table 1.3**

**Typical values of specific energy, specific power, and power flux density.**
$c = 3 \cdot 10^8$m/s (speed of light), $g = 9.81$ m/s$^2$ (terrestrial constant of gravitation).

becomes $\eta = \mathsf{U}/\mathsf{U}_0 = R/(R+R_0)$. You can then quickly show that the power $P = \mathsf{U}I = RI^2$ can be written as a function of $\eta$,

$$P(\eta) = \eta P_{tot} = P_0\eta(1-\eta) \qquad (1.2)$$

with $P_0 = \mathsf{U}_0^2/R_0$. The graph of the efficiency-power relation (1.2) is shown by the solid curve in Figure 1.1(b). You can observe three things. First, the power $P$ vanishes for $\eta \to 0$, which refers to vanishing load resistance $R \to 0$. This is understandable because the total power is then dissipated in the internal battery resistor, $R_0$. Secondly, the power also vanishes if $\eta \to \eta_{max} = 1$, because for constant voltage, the power $P \approx P_{tot} \approx \mathsf{U}_0^2/R$ decays as $1/R$ for $R \to \infty$. Thirdly, there exists a maximum power point $P_{max} = P_0/4$ at $\eta = 1/2$, where the resistances match, $R = R_0$.

Equation (1.2) does not contain leakage losses. If you like, you can include $R_L$ by using Kirchhoff's first law and subtract the current $\mathsf{U}_0/R_L$ leaking through $R_L$ from the current $I$ passing $R_0$ and $R$. It is then no longer possible to write a single function $P(\eta)$ or $\eta(P)$, but you can easily plot the efficiency-power relation as an implicit function parameterized by $R$ in the form $(\eta(R), P(R))$. In Figure 1.1(b), the numerically calculated result is shown by the dashed curve. The important consequence of leakage is an additional branch of vanishing efficiency at zero power, simply because infinite discharge time implies complete energy loss by leakage. You also see that the maximum efficiency, which appears now at finite power, is shifted to a value smaller

(a)                              (b)

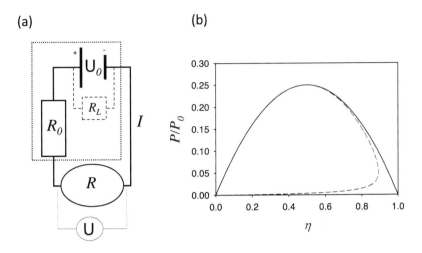

**Figure 1.1** (a) Electric circuit with an energy source (battery with voltage $U_0$, internal series resistance $R_0$, and leakage resistance $R_L$ within the dotted rectangle) and a consumer (load resistance $R$) at voltage $U$. (b) Efficiency-power relation of the system in (a) (solid: without leakage ($R_L \to \infty$); dashed: finite leakage $R_L$).

than one. More details of such efficiency-power relations will be discussed later, e.g., in Chapter 6. In particular you will also see then that Eq. (1.2) is not restricted to electrical networks, but is a prototype efficiency-power relation which describes various different applications. It can be generalized by a replacement of the voltage and current to generalized thermodynamic forces and currents, and also to time dependent cases.

   In the framework of the efficiency-power relations, the two general questions posed at the beginning of this section become more specific, and also more vivid. It is obvious that, for the purpose of optimizing energy devices or systems, a first task will be to *derive* the efficiency-power relation. When the efficiency-power relation is known, one has to decide on the optimum point on that curve. Obviously, important figures are the maximum values $P_{max}$ and $\eta_{max}$, respectively, of power and efficiency, because it is *the curve segment between these two points* that is relevant for applications. On this segment, an increase of one of the two objectives, efficiency and power, leads to a decrease of the other one, which is the manifestation of the trade-off problem. This curve segment should cover the operating range of an application.

## 1.3  OUTLINE

Chapter 2 starts with a quick review of equilibrium statistical thermodynamics in order to comprehend first the relevant terms and ideas linked to reversible processes. It should be clear that it is important to obtain first a deep understanding of the different forms of energy, and particularly of the special energy form *heat*. This chapter is probably the technically most difficult one for readers without much foreknowledge; but with enough fore-knowledge you may skim the chapter. In Chapter 3 linear nonequilibrium processes responsible for heat creation or exergy losses, like friction or leak-age, will be discussed. Nonequilibrium processes can be understood in many cases as generalized transport processes of extensive quantities driven by im-balances between intensive quantities, during which the entropy is increased. Chapter 4 is then devoted to the entropy production rate. After these basic ideas are covered, it is straightforward to describe general systems composed of reversible and irreversible parts. This will be done in Chapter 5 in the framework of endoreversible thermodynamics, where general systems are de-composed, roughly speaking, in reversible processes connected by irreversible links. Chapter 6 shows how energy storage devices can be described in terms of relations between power and harvested energy for constant power demand, so-called *Ragone-plots*. Chapter 7 discusses then a number of specific energy conversion applications and derives for them expressions for power and/or effi-ciency. Among them are solar, wind, hydro-power, as well as some important electrical and electro-mechanical conversion devices like transformers, con-verters, and motors. Additionally, *impedance matching*, which is important for maximizing power transfer, will be discussed especially for electrical appli-cations. The last chapter eventually focuses on optimization issues. An impor-tant question refers to the *optimum* efficiency-power relation of a composite system, like a hybrid of two devices with their individual efficiency-power relations. Another general question is connected to the optimum point on a given efficiency-power curve. Although some simple strategies, like power maximization, efficiency maximization, or entropy production minimization may be appropriate in certain limit cases, this issue is partly of an economic nature, and a general finalizing answer for technical devices can only be pro-vided by a techno-economic analysis. These topics will be touched on in the last chapter.

# 2 Equilibrium Statistical Thermodynamics

A solid understanding of the basic equilibrium thermodynamic notions like temperature, work, heat, entropy, etc., in the framework of statistical physics is a main prerequisite in order to comprehend nonequilibrium processes and irreversibilities, and efficiency and losses in energy conversion and storage. For this purpose, this chapter reviews some basics of statistical equilibrium thermodynamics. Although we start more or less from scratch, it is assumed that you once took a course in thermodynamics and statistics and will enjoy this chapter mainly as a refresher. The focus is on the fundamental concepts, with the cost of generality, completeness, and sometime even formal strictness. In case you want to learn more on statistical thermodynamics, there are many excellent textbooks about the matter (see, e.g., [Som96, Hil60]).

Let us start with recalling the four fundamental laws of thermodynamics, e.g., as formulated in the famous book by Arnold Sommerfeld [Som96]:

**Zeroth Law** All thermodynamic systems possess a property (state variable) called *(absolute) temperature*, $T$. Equality of the temperatures is a condition for thermal equilibrium between two systems or between two parts of a single system.

**First law** Part (i): All thermodynamic systems possess a property (state variable) called energy, $U$. Part (ii): The energy of a system is increased by the quantity of heat, $\delta Q$, absorbed by it, and by the external work, $\delta W$, performed on it. In an isolated system, the total amount of energy is preserved.

**Second law** Part (i): All thermodynamic systems possess a property called entropy, $S$. Part (ii): The entropy is calculated by imaging that the state of the system is changed from an arbitrary selected reference state to the actual state through a sequence of states of equilibrium and by summing up the quotients of the quantities of heat $\delta Q$, introduced at each step, and temperature $T$. During real processes the entropy of an isolated system increases.

**Third law** The entropy $S$ goes to zero if the absolute temperature $T$ goes to zero.

The third law, which implies that it is impossible to reach absolute zero temperature with a finite number of thermodynamic process steps, will be less relevant for our applications. Nevertheless, it is important for the purpose of understanding entropy and temperature. Later we will, for the same reason, even briefly show that the point $T = 0$ is not necessarily unique, as it can be reached in some systems from the positive and the negative temperature sides, and correspond to *different* states. If you want to know more about the third law, see Ref. [Wil61].

It is important to be aware that these basic laws *define* thermodynamic equilibrium. In this chapter, equilibrium thermodynamics is divided in two sections: *equilibrium states* and *reversible processes*. This division is in line with the above split of the first and the second laws into parts (i) and (ii). The discussion of equilibrium states only requires the introduction of the state variables $(T, U, S, ...)$, while the treatment of reversible processes involves state changes that are sequences of equilibrium states, which requires the additional introduction of process quantities like *heat* and *work*. Both sections elaborate the theory first and then provide a few illustrative examples.

## 2.1 EQUILIBRIUM STATES

Thermodynamics and statistical physics deal with macroscopic systems, like gases, liquids, solids, and plasma, which are composed of a huge number ($\sim 10^{23}$) of microscopic systems, like molecules, atoms, ions, electrons, photons, phonons, and other quasi particles. In the following, we will call the microscopic systems often *particles*. Their specific degrees of freedom like position, momentum, spin, etc., are generally functions of time and have values which must in principle be determined from the fundamental laws of physics, like from quantum mechanics or, in some cases, from classical approximations. Without dispute let us suppose that the most important property of a single, independent particle is its *energy*. Quantum mechanics tells us that the available states of such a particle, which is supposed to be in a possibly large but finite volume, can be numbered by $j = 1, 2, ...$ (sometimes, one starts with $j = 0$ for labeling the ground state with subscript 0) and ordered according to their (discrete) energy eigenvalues, $E_j$, with $E_{j+1} \geq E_j$. The task of statistical thermodynamics is to connect *thermodynamics*, which describes a macroscopic system by a few macroscopic, phenomenological variables, with the huge number of (micro-) states, which emerge from the composition of the many particles in the volume. For clarity, we shall first make a simplification by considering a system which consists of a fixed number $N$ of *independent, distinguishable classical particles*, and describe the way how to derive its equilibrium states by looking at an average particle. We shall then argue that equilibrium is related to maximum entropy. You might then ask: "*Who* maximizes entropy?" This is a serious question, and a serious answer is sophisticated and would go far beyond our purpose. For us it is sufficient to admit that the independent particles are *not* fully independent, in contrast to the above assumption. In fact, they *do* interact, i.e., exchange energy on a fast, microscopic time-scale such that thermodynamic *equilibration* can establish at all. But this must not be seen as an inconsistency: the effect of the interactions is so small that the energy values $E_j$ of the energy levels of the particles are not significantly affected. In a certain sense this interaction can be seen as a weak perturbation. Sometimes one finds in the literature the term *quasi-independent particles*, because one can treat them as independent. Since the weak microscopic interactions are responsible for the establishment of equilibrium, they will play an important role for irreversibilities and equilibration, which will be discussed in later chapters.

### 2.1.1  ENERGY

Let us suppose a system of $N$ quasi-independent particles in a finite volume $V$ which can be in states $i = 1, 2, ..., M$ having energies $E_i$ (we allow $M$ to be finite or infinite). We do not care how we found the energy levels but assume they are known. You may also wonder why they are discrete, yet quantum mechanics alone is not an argument, because even the hydrogen atom energy spectrum has not only a discrete (bound electron states) but also a continuous part (free electron states). However, in a finite volume $V$, the spectrum of a free electron becomes also discrete—because it is in truth not free but bound to the volume—although the spacings between the energy levels go to zero as $V$ becomes large and larger. The reason to insist on discreteness of the levels is because below we want to count them, and discreteness makes this easier. Of course, in order to count states from the set $\{E_j\}$, we need to know how many states belong to each energy level. Again recalling the H-atom, you know that several states can have the same energy (just remember the multiple quantum numbers for single atomic energy levels). Such *degeneracy* is usually due to the presence of symmetries and is lifted in cases of disturbances which break this symmetry, for example by weak interactions between the particles and/or small imperfections of their environment. The energy levels are then split in multiple levels (*multipletts*), as is well-known from basic atomic physics (e.g., fine-structure and hyper-fine-structure splitting). For our purposes it is at the moment sufficient (and convenient) to assume that every state is represented uniquely by one energy level, $E_i$, and the introduction of multiplicity factors $g_i$ for state counting is not necessary yet (we will do this later).

Let us now consider the average particle energy $U/N$, where $U$ is the total energy of the system of the $N$ quasi-independent particles. It is easy to do this if we know the distribution of the particles over the *microstates*. Assume that $n_i$ particles are in state $i$ for $i = 1, ..., M$. We thus define a probability distribution $w_i = n_i/N$ which tells us the probability of a particle of the system to be in state $i$. An illustration of the energy-level distribution is given in Figure 2.1. Note that the total number of states $M$ is not necessarily large or infinite, it can be small - while we assume the total number of particles, $N$, to be large. The average energy $U/N$ can thus be written as,

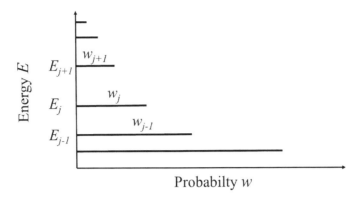

**Figure 2.1** Energy levels with energy $E_j$ and probabilities $w_j$.

$$\frac{U}{N} = \sum_{i=1}^{M} w_i E_i.$$  (2.1)

Normalization of the *occupation probabilities* $w_i$ means

$$\sum_{i=1}^{M} w_i = 1.$$  (2.2)

We know the $E_i$-values. However, we also need the $w_i$-values for determining $U$. In general, you can have an arbitrary $w_i$-distribution, but in this chapter we are interested in the *thermodynamic equilibrium distribution*.

## 2.1.2  ENTROPY MAXIMIZATION

Since $N$ is large, we use statistics for finding the distribution with the highest probability, i.e., with the largest number of microstates (i.e., realizations of a macrostate), which are in accordance with the given constraints that define the macrostate. The resulting distribution $\{w_i^{(eq)}\}_{i=1,2,\ldots}$ is the so-called *thermodynamic equilibrium distribution*. We must thus find the maximum number of ways to distribute the $N$ distinguishable particles to the $M$ energy levels. Basic combinatorics yields the number of ways to realize the distribution with $n_1$ particles on level $E_1$, $n_2$ particles on level $E_2$, ..., $n_M$ particles on level $E_M$:

$$\frac{N!}{n_1! n_2! \ldots n_M!}.$$  (2.3)

It is more convenient to maximize the logarithm of (2.3), which is equivalent to maximizing (2.3) itself since $\ln(x)$ is a strictly monotonous function. It is convenient because the logarithm of a product transforms into a sum of logarithms, such that (2.3) becomes

$$\ln(N!) - \sum_{i=1}^{M} \ln(n_i!),  \tag{2.4}$$

and because for large numbers $x$ Stirling's approximation $ln(x!) \approx x(\ln(x) - 1) = x \ln(x/e)$ helps to simplify the final expression (2.3), which then reads

$$\ln(N!) - \sum_{i=1}^{M} n_i \ln(n_i) + \sum_{i=1}^{M} n_i.  \tag{2.5}$$

We will not discuss the assumption of all $n_i$ being large here, which is not always true, but is a prerequisite for the Stirling approximation (here is now a point where we oversimplify statistical physics, but this is irrelevant for *our* purpose). Since the logarithm of (2.3) is used only for maximization and is a function of the distribution $n_i$ (or $w_i$), the addend $\ln(N!)$ in the sum, which does not depend on any $w_i$, can be disregarded. Similarly the last term drops out, because the sum of the $n_i$ gives just $N$. What remains is to maximize the sum of $-Nw_i \ln(Nw_i) = -Nw_i(\ln w_i + \ln N)$. The $\ln(N)$-term is again irrelevant since the sum $\Sigma w_i \ln N$ is constant. Therefore we are left with the task to maximize the expression

$$S(w_i) = -k_B N \sum_{i=1}^{M} w_i \ln w_i,  \tag{2.6}$$

where we added an arbitrary constant prefactor $k_B$. As you certainly know, in thermodynamics $k_B$ is usually the Boltzmann constant, $k_B = 1.38\,10^{-23}\,J/K$. Up to some irrelevant differences, the expression (2.6) is also known as *Shannon entropy* in information theory. It is important to be aware of that the entropy (2.6) is *a purely statistical quantity* and is not restricted to equilibrium states alone, while the definition of the *thermodynamic* entropy, as will become obvious later, *is* restricted to equilibrium states where it is consistent with the definition Eq. (2.6).

Obviously, $S = 0$ in case of certainty, i.e. if a single $w_i$ equals 1 and all other $w_k$ ($k \neq i$) vanish. Otherwise $S$ is positive. In the case of largest uncertainty, where all states $i$ have the same occupation probability $w_i = 1/M$ for all $i$ (*equipartition*), the entropy (2.6) is

$$S = k_B \ln(M),  \tag{2.7}$$

which is the famous entropy expression of Boltzmann (he used the symbol $\Omega$ for the total number of states (*volume of phase space*). *Certainty* and *uncertainty* are sometimes called, respectively, *order* and *disorder* for obvious reason.

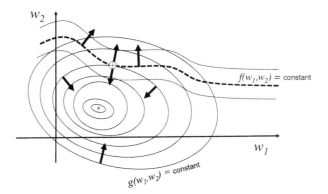

**Figure 2.2** In order to find the optimum ($\oplus$) of a function $g(\{w_k\})$ (solid closed loops) subject to a constraint $f(\{w_k\}) = 0$ (thick dashed curve) one has to search for their point of contact, i.e., where the gradient vectors (arrows) of $f$ and $g$ are collinear: $\nabla f \propto \nabla g$.

Equipartition holds if all attainable energy levels are so close that they can be considered equal (an energy constraint is then obsolete). In practice, it means that the occupied energy levels are centered at a fixed energy and lying in a range $\Delta E \ll k_B T$. This is called the *microcanonical ensemble*. However, we want to find the equilibrium distribution $w_i$, $i = 1, 2, ...$, for the *canonical ensemble* where levels with different energies can be occupied, and the average energy is prescribed by a constraint. Hence, $S(\{w_i\})$ must be maximized with both the constraints given by (2.1) and (2.2) for given values of $U$ and set $\{E_i\}$.

Most of you will know that optimization with constraints leads to the introduction of Lagrange multipliers. For those who do not know (or forgot), there is a simple picture to understand how it goes. The Lagrange multipliers have, in thermodynamics, an important meaning: they manifest themselves as temperature, chemical potential, pressure, and other so-called *intensive* quantities, as you will see below. Later we will encounter further optimization problems where we will apply the approach.

As an analogy, imagine you trace a path in a hilly landscape with plane coordinates $w_1$ and $w_2$ (i.e., here $M = 2$), on a map which contains altitude curves with $g(w_1, w_2) =$ constant. The path is described by a function $f(w_1, w_2) = 0$, and you ask for the maximum altitude along that path. If you make a picture, as is done in Figure 2.2, you immediately see that at the point of highest altitude, the normal vectors of the path $f$ and of the altitude curve $g$ have the same direction (they are collinear). This means that the gradients of the functions defining the altitude and the path, are proportional, $\partial g / \partial w_k = \lambda \partial f / \partial w_k$, for all $k$. The proportionality constant $\lambda$ is the so-called *Langrange multiplier*. Since its value is unknown, it increases the number of unknowns by one, but there is no overdeterminacy because the constraint $f = 0$ serves as the additional equation required for having equal number of unknowns and equations. Of course, in case of doubts we must ensure that the optimum is a maximum by looking at the second derivative along the path, and if there are several local maxima, we have to compare them for finding the global maximum.

For entropy maximization, the variables are not just two coordinates but $M$ probabilities, $w_i$, and there are two (or more) constraints. Here, they are expressed as $f_0 = \Sigma w_i - 1 = 0$ (normalization) and $f_1 = \Sigma E_i w_i - U/N = 0$ (average energy). After introducing the two Lagrange multipliers $\alpha - 1$ and $\beta$, maximization of $f = S/Nk_B$ with the mentioned constraints gives

$$\frac{\partial}{\partial w_k}\left(\Sigma_1^M w_i \ln(w_i) + (\alpha - 1)(\Sigma_1^M w_i - 1) + \beta(\Sigma_1^M w_i E_i - U/N)\right) = 0, \quad (2.8)$$

for $k = 1, ..., M$, which leads to

$$\ln w_i + \alpha + \beta E_i = 0. \quad (2.9)$$

The two unknowns $\beta$ and $\alpha$ are determined with the two Eqs. (2.1) and (2.2). Equation (2.9) gives then the Boltzmann distribution

$$w_i^{(eq)} = \exp(-\alpha - \beta E_i) = \frac{\exp(-\beta E_i)}{Z}, \quad (2.10)$$

for the single particle, where the superscript $(eq)$ indicates thermodynamic equilibrium, and $Z = \exp(\alpha)$ is the normalization constant

$$Z = \sum_{i=1}^{M} \exp(-\beta E_i), \quad (2.11)$$

known as *partition function* for the single particle. It is the *sum of Boltzmann factors* $\exp(-\beta E_i)$ *over all microstates* $i$. If you have a system with degenerate states of multiplicity $g_i$ and want to count every energy level only once, you can also write

$$Z = \sum_{i=1}^{M} g_i \exp(-\beta E_i). \quad (2.12)$$

This will be convenient particularly when the sum over $i$ becomes an energy integral over $E$, and $g(E)$ is the density of states.

By inserting the equilibrium distribution (2.10) in the general entropy expression Eq. (2.6), the entropy can be written as

$$S = Nk_B \sum_{i=1}^{N} w_i \left(\beta E_i + \ln(Z)\right) = k_B \beta U + k_B \ln(Z_N), \quad (2.13)$$

which will turn out to be a very useful relation. The second term on the right-hand side of Eq. (2.13) contains the $N$-particle partition function, which reads obviously $Z_N = Z^N$. Unfortunately, it turns out that this is not fully correct. Quantum mechanics teaches that identical particles are indistinguishable. For instance, a permutation of two identical particles, one in single-particle state 1 and the other in state 2, does not lead to another new state where the first one is now in state 2 and the second one in state 1. It is just a single state, to have one in 1 and one in 2, because it makes no sense to label quantum mechanical particles by "first one" and "second one." Therefore, in order to count the $N$-particle states correctly and not to end up with too many, one must divide the classical result by the number of possible permutations, such that

$$Z_N = \frac{Z^N}{N!}, \tag{2.14}$$

is the correct $N$-particle partition function in Eq. (2.13). This is a situation where already for a classical quantity a quantum effect must be taken into account. But because we considered here $N$ as a fixed number, this correction leads only to a constant term in the entropy without any effect in the classical regime. Partition functions are often in the center of statistical thermodynamics, because all equilibrium properties can be derived from them. For example, the average energy can be obtained from

$$U = -\frac{\partial \ln(Z_N)}{\partial \beta}, \tag{2.15}$$

which follows immediately from Eqs. (2.1), (2.10), and (2.11). The mentioned constant factor $1/N!$ just drops out. However, at low temperatures and in cases where $N$ is not fixed you have to take it into account. Prior to a brief discussion of quantum statistics, we show that the Lagrange multiplier $\beta$ is related to the temperature, which is - remember the zeroth law - the important state variable associated with thermodynamic equilibrium.

### 2.1.3  TEMPERATURE

Assume a thermodynamic equilibrium system defined by a given energy spectrum $\{E_i\}$ and probabilities $\{w_i^{(eq)}\}$. Note that Eq. (2.11) implies that $Z$ (or $Z_N$) is $U$-independent if $\beta$ and the $E_i$ are constant. If one considers $S$ as a function of $U$, Equation (2.13) implies then $dS/dU = k_B\beta$, and with the *Kelvin temperature* defined by $T = 1/k_B\beta$, one obtains

$$\frac{1}{T} = \frac{\partial S}{\partial U}. \tag{2.16}$$

We used here partial derivatives in order to anticipate that later more state variables may be involved (like the volume, particle number, etc.), which are kept fixed at the moment (because the energy levels are fixed by assumption).

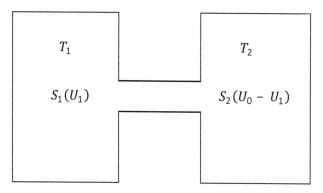

**Figure 2.3** Isolated system with energy $U_0$ and consisting of two subsystems, $k = 1, 2$, with energies, entropies, and temperatures $U_k$, $S_k$, and $T_k$, respectively

Equation (2.16) reads in words: *the change in the entropy (disorder) per energy change is a measure for the inverse temperature.* For large $T$ (*hot* system) the entropy increase is small when the energy is increased.

Why $T$ is a temperature according to the zeroth law of thermodynamics can be made clear now. Consider for this purpose Figure 2.3 with two systems 1 and 2 which are weakly coupled, such that energy exchange is possible ($U_1$ and $U_2$ may change), but the coupling is so weak that all so-called *extensive quantities*, like $V$, $U$, $S$, just add. Thermal equilibrium means that the total entropy, $S = S_1(U_1) + S_2(U_2)$ is maximum. You can write $U_2 = U_0 - U_1$ for the energy of the second system, since the total system is supposed to be isolated and has thus constant energy $U_0$. We may consider $U_1$ as the free variable, and to find its equilibrium value we must maximize the total entropy $S(U_1) = S_1(U_1) + S_2(U_0 - U_1)$. Hence

$$\left(\frac{\partial S_1}{\partial U}\right)_{U=U_1} - \left(\frac{\partial S_2}{\partial U}\right)_{U=U_2} = 0 \tag{2.17}$$

which means with Eq. (2.16) that

$$T_1 = T_2. \tag{2.18}$$

If the two systems are in thermodynamic equilibrium, they have equal temperatures; that is the main part of the zeroth law. The just discussed approach with Figure 2.3 will later become important in a generalized form in the context of nonequilibrium and current flows. By the way, although the mentioned issue is usually illustrated by two separated reservoirs as in Figure 2.3, the two systems are not necessarily *spatially* separated; the two reservoirs may also describe, for example, different particle species which are mixed in space, like electrons and ions in a plasma, or different degrees of freedom of molecules in a lattice, like magnetic moments and vibrations.

## 2.1.4   STATISTICS OF QUANTUM PARTICLES

In the previous section, the underlying picture of a statistical ensemble was the following. We considered just an *average* single particle in the volume $V$ with fixed number $N$ of particles. *Average* may be understood as having many virtual copies of it, and then we do statistics with this *ensemble*. We then ended up with the single-particle partition function (2.11), and took (2.14) for the $N$-particle system. Without going deeply into details, we emphasize that different types of ensembles can be related to different setups of constraints to the entropy maximization problem. It was already mentioned in connection with Eq. (2.7) that if the entropy is maximized under the single constraint (2.2) for normalization, the ensemble is called *microcanonical*. And if additionally the constraint (2.1) for the average energy is taken into account, the ensemble is called *canonical ensemble*. When it comes to quantum mechanical particles (fermions or bosons), the most convenient ensemble is the so-called *grand-canonical ensemble*. The states are then characterized not solely by energy for one fixed particle number $N$ for the total system, but also by different particle numbers $N_i$. In other words, systems with different energies and particle numbers now build the set of the statistical ensemble.

You might wonder, why quantum statistics is discussed in this brief review. The reason is that in Section 7.1 we shall try to understand solar energy conversion, which requires knowledge about the thermodynamics and the distribution function of photons — and the number of photons is not a conserved quantity in general. Furthermore, we will then encounter not only bosons but also fermions, namely, electrons and holes in conductors and photodiodes. Nevertheless, we still remain again concise and focus on the most important points. If you like, you may skip part of this section, and just have a look at the main results, which are Eqs. (2.28) and (2.30).

Quantum particles of the same type are *indistinguishable*, which means (as quickly mentioned earlier) that the multiplicity of one realization of a particle distribution, $\{n_i\}$, equals one, because a permutation which leads to the same $\{n_i\}$ gives the *identically same* state. A specific realization of a particle distribution $\{n_i\}$ refers to: $n_1$ particles are in state 1, $n_2$ are in state 2, ... . A given distribution $\{n_i\}$ is thus characterized by a set of numbers $(n_1, n_2, n_3, ..., n_M)$. The total particle number $N$ of a state with this $\{n_i\}$ is

$$N(\{n_i\}) = \sum_{i=1}^{M} n_i,\tag{2.19}$$

and the associated energy is

$$U(\{n_i\}) = \sum_{i=1}^{M} n_i E_i.\tag{2.20}$$

It should be obvious from formal analogy to what we did before, that the maximization of the entropy leads to probabilities with Boltzmann factors of the form

$$w^{(eq)}(\{n_i\}) = \frac{\exp(-\beta[U(\{n_i\}) - \mu N(\{n_i\})])}{Z_G}. \qquad (2.21)$$

The additional constraint (2.19) led to an additional Lagrange multiplier, which we write as $-\beta\mu$. The parameter $\mu$ is the *chemical potential*. The normalization constant,

$$Z_G = \sum_{\{n_i\}} \exp\left(-\beta \sum_i (E_i - \mu)n_i\right), \qquad (2.22)$$

is called the *grand-canonical partition function*. The outer sum runs over all possible distributions $\{n_i\}$, while the sum in the exponent runs over all $i$ for a fixed distribution $\{n_i\}$. Because $U(\{n_i\})$ and $N(\{n_i\})$ for a given realization $\{n_i\}$ are sums, the $w^{(eq)}$ in Eq. (2.21) can be written as products

$$w^{(eq)}(\{n_i\}) = \frac{\exp(-\beta \sum_i (E_i - \mu)n_i)}{Z_G} = \prod_i \frac{\exp(-\beta(E_i - \mu)n_i)}{Z_i} = \prod_i w_i^{(eq)}. \qquad (2.23)$$

For the partition function, we used that for the product of sums it holds

$$\prod_i \left(\sum_{n_i} h_i(n_i)\right) = \sum_{n_1}\sum_{n_2} \cdots \sum_{\cdots} \cdots \prod_i h_i(n_i) = \sum_{\{n_i\}} \prod_i h_i(n_i). \qquad (2.24)$$

Equation (2.23) is an important interim result: for these quasi-independent particles the probabilities factorize, i.e., the different occupation numbers $n_i$ are statistically independent. One can thus define a probability distribution for the occupation $n_i$ of each level $i$,

$$w_i^{(eq)} = Z_i^{-1} \exp(-\beta(E_i - \mu)n_i), \qquad (2.25)$$

with normalization constants

$$Z_i = \sum_{n_i} \exp(-\beta(E_i - \mu)n_i). \qquad (2.26)$$

Because the $Z_i$ are the partition functions of subsystems associated with a disjoint and complete decomposition of the total system, the total partition function is the product of the single-level partition functions, $Z_G = \prod_i Z_i$.

If you wonder why we did not specify in the sum (2.26) the set of numbers $n_i$ (for instance, from $n_i = 0$ to $\infty$), here is the simple answer: it depends on whether the particles are bosons or fermions. Let us first consider fermions (like electrons) where, according to the Pauli principle, a level can at most be occupied with one particle (or two, if the level has spin degeneracy). Hence, $n_i = 0$ and $n_i = 1$ are the only possibilities, which leads to a $Z_i$ given by

$$Z_i^F = 1 + \exp(-\beta(E_i - \mu)). \qquad (2.27)$$

The average occupation number $f(E_i)$ of level $i$ is then given by

$$f(E_i) = \sum_{n_i=0}^{1} n_i w_i^{(eq)} = \frac{1}{\exp(\beta(E_i - \mu)) + 1}, \tag{2.28}$$

the *Fermi−Dirac distribution*, which is in many relevant cases more or less a step function which goes from $f = 1$ to $f = 0$ at the chemical potential $\mu$ (which is also called *Fermi energy* in this context) in an energy interval with a width of the order of the thermal energy $\beta^{-1} = k_B T$ (see Figure 2.4(a)).

Bosons, like photons and phonons, on the other hand, can take all $n_i$ values $0, 1, 2, ....$ The partition function is thus a geometrical series,

$$Z_i^B = \sum_{n_i=0}^{\infty} \exp(-\beta(E_i - \mu)n_i) = \frac{1}{1 - \exp(-\beta(E_i - \mu))}. \tag{2.29}$$

If one defines as usual the energy scale such that the ground state has zero energy, the chemical potential $\mu$ of bosons must be negative, $\mu < 0$. Otherwise the partition function diverges. There are two exceptions. First, mass-less bosons like photons cannot have energies that are identically zero. Because this means $E_i > 0$, the chemical potential must satisfy only $\mu \le 0$, which includes the equality. It will turn out that the equilibrium photon gas has $\mu = 0$. Secondly, there are some special cases where $\mu \to 0$, like Bose condensation (e.g., super-fluidity of He), but this will not be considered here.

The average occupation number for bosons is given by

$$f(E_i) = \sum_{n_i=1}^{\infty} n_i w_i^{(eq)} = \frac{d \ln(Z_i^B)}{\beta \, d\mu} = \frac{1}{\exp(\beta(E_i - \mu)) - 1}, \tag{2.30}$$

which is the well-known *Bose−Einstein distribution* function (see Figure 2.4).

For both Fermi−Dirac and Bose−Einstein distributions it holds that

$$N = \sum_{i=1}^{M} f(E_i) \tag{2.31}$$

and

$$U = \sum_{i=1}^{M} f(E_i) E_i. \tag{2.32}$$

Because the levels can be seen as independent subsystems, one may define entropies for each single energy level $i$ by

$$S_i(f_i) = -k_B \sum_{n_i} w_i \ln w_i \tag{2.33}$$

by using Eq. (2.25) for the $w_i$. The total entropy is the sum of all single level entropies,

$$S = \sum_{i} S_i. \tag{2.34}$$

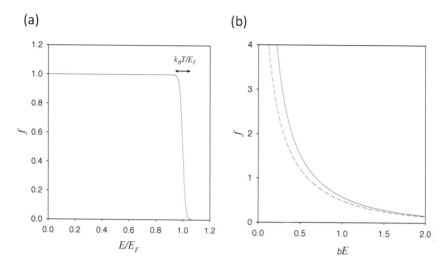

**Figure 2.4** (a) Fermi–Dirac distribution $f$ as a function of $E/E_F$, with chemical potential $\mu = E_F$ (Fermi-energy). (b) Bose–Einstein distribution $f$ as a function of $\beta E$ for finite (dashed) and vanishing (solid) chemical potential.

For later use, when we have to investigate the entropy current of heat radiation, we determine $S_i$ for bosons. A combination of (2.25), (2.29), and (2.30) yields

$$w_i^{(eq)} = \frac{1}{1 + f_i} \left( \frac{f_i}{1 + f_i} \right)^{n_i}, \tag{2.35}$$

and thus

$$\ln(w_i^{(eq)}) = -\ln(1 + f_i) + n_i \left( \ln(f_i) - \ln(1 + f_i) \right). \tag{2.36}$$

Substitution in Eq. (2.33) and using Eq. (2.28) gives for the entropy of level $i$

$$S_i(f_i) = -k_B \left( f_i \ln(f_i) - (1 + f_i) \ln(1 + f_i) \right). \tag{2.37}$$

In a similar way, the single-level entropies for fermions can be derived,

$$S_i(f_i) = -k_B \left( f_i \ln(f_i) + (1 - f_i) \ln(1 - f_i) \right). \tag{2.38}$$

On the other hand, from Eqs. (2.21) and (2.22) one obtains for the entropy in the same way as Eq. (2.13)

$$S = -k_B \sum_{\{n_i\}} w^{(eq)} \ln(w^{(eq)}) = k_B \beta (U - \mu N) + k_B \ln(Z_G). \tag{2.39}$$

Equations (2.37) and (2.38) can be understood as general entropy expressions for nonequilibrium distributions $f_i$ [LL13] (although in nonequilibrium, the definition of an entropy is not necessarily *unique*).

### 2.1.5  EXAMPLES OF EQUILIBRIUM STATES

The previous subsections aimed to be intuitively accessible, but were still rather formal. For practical illustration and understanding three examples will be discussed next. The first one concerns results on the ideal gas that you have seen already elsewhere, the second one refers to the photon gas and is relevant for solar power, and the third one on negative absolute temperatures aims to give a deeper understanding of the notions *temperature* and *entropy*.

#### 2.1.5.1  Perfect gas of independent particles

Quasi-independent particles with mass $m$ in a volume $V$ should be treated quantum-mechanically or at least semi-classically, if one calculates the $N$-particle partition function $Z_N$. The reason is the non-discreteness of the classical energy spectrum. The kinetic energy of a single particle is

$$E = \frac{\sum_{k=1}^{3} p_k^2}{2m}, \tag{2.40}$$

where the $p_k$'s are the continuous momentum coordinates in the three space directions. Of course, if you just want to know the probability distribution function $f$ in momentum or velocity $(v_i = p_i/m)$ space, you can determine the normalization constant $\mathcal{N}$ of $f(\vec{v}) = \mathcal{N} \exp(-\beta m \vec{v}^2/2)$. This will lead to the well-known Maxwell velocity distribution - however we will not follow this path, but aim for determining $Z_N$. It is not clear, without quantum mechanics, how to count energy states for these continuous variables. Heisenberg's uncertainty relation

$$\Delta x_k \Delta p_k \geq h, \tag{2.41}$$

where $h = 6.626 \cdot 10^{-34} Js$ is the Planck constant, is sufficient to do this. Equation (2.41) holds for each of the three coordinate directions, and means that a single, one-dimensional particle state requires at least a phase-space $(x\text{-}p)$ area of size $h$. In other words, in an area $dx\,dp$ there are $dx\,dp/h$ states. For $N$ particles in three space dimensions, the phase-space volume has $3N$ dimensions, such that the $N$-particle partition function (2.14) becomes

$$Z_N = \frac{V^N}{N! h^{3N}} \left( \int_{-\infty}^{\infty} dp \, \exp\left(-\beta \frac{p^2}{2m}\right) \right)^{3N}, \tag{2.42}$$

with the above-mentioned $N!$ in the denominator. We also made use of the assumption that the particle energy in the volume is independent of space coordinates, which leads to the factor $V$ when integrating over $dx_1 dx_2 dx_3$ for a single particle. The $p$-integral in Eq. (2.42) can be calculated from the Gauss integral

$$\int_{-\infty}^{\infty} ds \, \exp(-s^2) = \sqrt{\pi}, \tag{2.43}$$

by changing to the integration variable $s = p\sqrt{\beta/2m}$. One obtains

$$Z_N = \frac{V^N}{N!}\left(\sqrt{\frac{2\pi m}{\beta h^2}}\right)^{3N} = \frac{1}{N!}\left(\frac{V}{\lambda_{th}^3}\right)^N, \qquad (2.44)$$

for the partition function. Here, the so-called *thermal de Broglie wavelength*

$$\lambda_{th} = \sqrt{\frac{h^2\beta}{2\pi m}} = \frac{h}{\sqrt{2\pi m k_B T}} \qquad (2.45)$$

is introduced as an abbreviation, and refers to the quantum mechanical wavelength of a particle having the thermal energy associated with the temperature $T$. To see this, you should recall the momentum-wavelength relationship, $\lambda = h/p$, from basic quantum mechanics, and if you combine it with a particle having kinetic energy $p^2/2m \approx k_B T$ (where a factor $\pi$ turns out to be here the correct coefficient of $k_B T$) you will end up with something like Eq. (2.45). The volume required by a particle is thus $\lambda_{th}^3$. Because matter particles usually show fermionic behavior, only one particle (for a given spin) can occupy this volume $\lambda_{th}^3$. There are $Z = V/\lambda_{th}^3$ volumes of size $\lambda_{th}^3$ contained in $V$. $Z$ is thus the single-particle partition function, and we know now how to get from it the $N$-particle partition function $Z_N$ by division of $N!$. That is what is behind Eq. (2.44). With Eq. (2.15) you immediately get the energy

$$U = -\frac{\partial \ln(Z_N)}{\partial \beta} = \frac{\partial}{\partial \beta}\ln(\beta^{3N/2}) = \frac{3N}{2\beta} = \frac{3}{2}Nk_B T. \qquad (2.46)$$

You certainly know this result: it is the *caloric equation of state* for the perfect gas. The energy of the perfect gas depends only on the temperature and the particle number, and each of the $3N$ degrees of freedom (here the momentum) shares the same amount of energy, $k_B T/2$ (*equipartition theorem*). If the number $N$ is fixed, $U$ is independent of the volume $V$ and depends only on $T$.

The entropy is obtained from Eq. (2.13):

$$S = \frac{3}{2}Nk_B + k_B\ln(Z_N). \qquad (2.47)$$

The second term on the right-hand side becomes, with the Stirling approximation $\ln(N!) \approx N\ln(N/e)$,

$$\ln(Z_N) = \ln\left(\frac{V^N}{(N/e)^N\lambda_{th}^{3N}}\right), \qquad (2.48)$$

such that

$$S = Nk_B\left(\frac{3}{2} + \ln(\frac{eV}{N\lambda_{th}^3})\right) = Nk_B\left(\frac{5}{2} + \ln(\frac{V}{N\lambda_{th}^3})\right), \qquad (2.49)$$

which is known as the *Sackur–Tetrode* formula. With the Eqs. (2.46) and (2.49), we have now the thermodynamic equilibrium relations between the state variables $T$, $U$, $V$, and $S$.

## 2.1.5.2 The photon gas

We will see in Section 4.2.2 that heat radiation is very special because photons do not interact with each other. Even not a little bit as would be required, as earlier mentioned, for the establishment of thermodynamic equilibrium. Consequently, the photon gas alone cannot equilibrate. It needs the interaction with something else, namely, with matter. Therefore, the photon gas at equilibrium is usually described as *cavity radiation (Hohlraumstrahlung)*, which exists in an otherwise empty volume enclosed by solid walls that are in equilibrium with the radiation. Of course, the walls themselves must be in thermodynamic equilibrium at a temperature $T$, and there should be an interaction between matter and photons of all wavelengths, which allows equilibration of the photons at all wavelengths to the unique temperature $T$. This is the case for so-called *blackbodies*, which have a surface that absorbs all incoming radiation and emits equilibrium radiation at temperature $T$. Further on, when solar and photovoltaic power will be discussed, we will come back to that. Here we just introduce the equilibrium state of the photon gas by using the Bose−Einstein distribution (or, because $\mu = 0$, the *Planck distribution*), and will postpone the topic of heat radiation to Section 3.3.2.

Photons are *massless* bosons. Note that from Eq. (2.21) one may interpret the chemical potential $\mu$ as the energy cost of adding one additional particle, independent of the energy level $j$. The energy states of photons with wavenumber $\vec{k} = (k_1, k_2, k_3)$ have an angular frequency $\omega = ck = c \mid \vec{k} \mid$, where $c$ is the speed of light, and thus have energy

$$E = \hbar\omega = \hbar ck. \tag{2.50}$$

In contrast to particles with finite rest mass $m_0$, that have finite energy $E = m_0 c^2$ even in the lowest classical energy state with zero momentum (or wavenumber), the photon energy exactly vanishes in the low-frequency limit $\omega \to 0$. It means that the energy cost becomes arbitrarily small for adding a photon with a sufficiently long wavelength to the photon gas. This suggests $\mu = 0$ for the equilibrium photon-gas. We anticipate here that this can be different if there is a low-energy cut-off, as can be the case in photovoltaic devices when the energy spectrum of the matter that equilibrates the radiation, has an energy gap. But that is nonequilibrium physics and appears in later chapters.

Since $\mu = 0$, the photon occupation function (2.30) becomes

$$f = \frac{1}{\exp(\beta\hbar ck) - 1} = \frac{1}{\exp(\beta\hbar\omega) - 1}. \tag{2.51}$$

This distribution function can be traced back to the famous Planck radiation law, which was introduced before the Bose−Einstein distribution. We give in Eq. (2.51) both, $k$ and $\omega$, representations. Angular frequency $\omega$ is more usual, while the wave number $\vec{k}$ is more useful, since for the calculation of non-isotropic quantities (like the heat flow density), one needs to integrate over sub-regions of $\vec{k}$-space, or spatial directions. With Eq. (2.41), and $\Delta p = \hbar \Delta k$, one has one state in the area $\Delta x \Delta k = 2\pi$. Since a photon can have two polarization directions in the plane normal to $\vec{k}$, there is a factor of 2 for the density of states, which leads to $2/(2\pi)^3$ in three dimensions. Hence the photon number of states in the six-dimensional volume element $d^3x\, d^3k$ is $2d^3x\, d^3k/(2\pi)^3$. The total number of photons and the total energy of the photon gas in $V$ become thus, respectively,

$$N = \frac{2V}{(2\pi)^3} \int d^3k f(\vec{k}) = \frac{V}{\pi^2 c^3} \int_{\omega=0}^{\infty} \frac{\omega^2 d\omega}{\exp(\beta\hbar\omega) - 1}, \qquad (2.52)$$

and

$$U = \frac{2V}{(2\pi)^3} \int d^3k \hbar c k f(\vec{k}) = \frac{V}{\pi^2 c^3} \int_{\omega=0}^{\infty} \frac{\hbar\omega^3 d\omega}{\exp(\beta\hbar\omega) - 1}, \qquad (2.53)$$

where we used for the second equality the isotropy of the integrand, and that $d^3k \to 4\pi k^2 dk = 4\pi\omega^2 d\omega/c^3$ after integration over the solid angle. The energy $U$ will be evaluated further. Introducing the integration variable $s = \beta\hbar\omega$, (2.53) becomes

$$U = \frac{V}{\pi^2 (\hbar c)^3 \beta^4} \int_{s=0}^{\infty} \frac{s^3 ds}{\exp(s) - 1} = aT^4 V, \qquad (2.54)$$

where the *Stefan constant* $a = \pi^2 k_B^4/15(\hbar c)^3$ has the value $7.565 \cdot 10^{-6}$ J/m$^3$K$^4$ (the definite integral over $s$, which equals $\pi^4/15$, can be found in appropriate textbooks or mathematical tables). The entropy follows then from Eq. (2.39) with $\mu = 0$:

$$S = k_B \beta U + k_B \ln Z_G. \qquad (2.55)$$

We need to evaluate the last term. Since the total partition function $Z_G$ is the product of all single-level partition functions $Z_i^B$, one obtains with Eq. (2.29) for $\mu = 0$

$$\ln(Z_G) = \ln\left(\prod_\omega \frac{1}{1 - \exp(-\beta\hbar\omega)}\right) = -\sum_\omega \ln(1 - \exp(-\beta\hbar\omega)). \qquad (2.56)$$

The sum runs over all mode frequencies $\omega$. We turn this sum into an integral over frequency, taking the above used density of states into account,

$$\sum_\omega \to \frac{V}{\pi^2 c^3} \int_0^{\infty} \omega^2\, d\omega, \qquad (2.57)$$

such that $\ln(Z_G)$ can be written as

$$-\frac{V}{\pi^2 c^3} \int_0^{\infty} \omega^2 d\omega \ln(1 - \exp(-\beta\hbar\omega)) = \frac{V}{\pi^2 c^3} \int_0^{\infty} \frac{\beta\hbar\omega^3 d\omega}{3(\exp(\beta\hbar\omega) - 1)}, \qquad (2.58)$$

where the right-hand side follows from integration by parts. Comparison with Eq. (2.53) shows that the last integral is just $\beta U/3$, such that Eq. (2.55) gives the entropy of the photon gas:

$$S = \frac{4k_B \beta U}{3} = \frac{4U}{3T} = \frac{4}{3}aT^3 V. \tag{2.59}$$

As for the perfect particle gas, we now have the relations between $T$, $V$, $U$, and $S$ of the equilibrium photon gas. You might wonder why Eq. (2.16) seems to be violated by Eq. (2.59). Of course, it is *not* violated; just recall that we emphasized the use of *partial* instead of *total* derivatives, and made the assumption that other variables, like $N$, are kept constant. For massive particles, we can fix $N$, but not for the photons: $N$ will change according to Eq. (2.52) when the energy changes.

### 2.1.5.3   Two-level system and negative temperatures

As mentioned, one purpose of this section is to broaden our horizon with respect to the notion of *temperature*. The discussion of systems where the single-particle energy-spectrum exhibits a finite number $M$ of energy levels leads to the somewhat exotic concept of *negative absolute temperatures*. To understand this is easy and of didactical value. The results will also indicate that - although in practice it is commonly $T$ that is used as a measure for cold and hot - for fundamental theoretical discussions it can be more natural to use the quantity $-\beta$ for hotness. A deeper discussion of the concept of negative temperatures, including experimental validation, can be found in standard textbooks on thermodynamics (see, e.g., [Wil61]). For us it is sufficient to illustrate it for two-level systems, where things become especially simple.

Most systems you know have energy levels which are bounded from below but unbounded from above: the energy can become arbitrarily large. Just recall our two previous examples. Without saying, we supposed therefore in the previous sections that $\beta$ was positive, in order for the partition function $Z$, Eq. (2.11), to remain finite. Negative $\beta$ (or negative absolute temperature $T < 0$) values were excluded. However, if the partition sum $Z$ has a *finite* number $M$ of terms, a maximum energy $E_M$ exists, and $\beta$ can be negative without causing any troubles. Let us consider the simplest example of a nontrivial thermodynamic system you could imagine. It consists of particles with only two energy levels, say $E_1 = 0$ (ground state) and $E_2 = E$ (excited state). We then have a single independent probability, say $w := w_2 = 1 - w_1$ for the upper level. For instance, a localized (i.e., fixed on a lattice) spin $1/2$ in a magnetic field behaves like this (the energy would then be the product of the magnetic field and the magnetic moment with only two directions). The mean energy per particle is

$$\frac{U}{N} = wE. \tag{2.60}$$

This case is so simple because the equilibrium distribution, $1 - w$ and $w$, is completely determined if $U/N$ is given. Equation (2.10) reads

$$w = \frac{\exp(-\beta E)}{1 + \exp(-\beta E)} \tag{2.61}$$

which implies, that $U/N$ is also a well-defined function of $\beta$. The same holds for the entropy,

$$S = -Nk_B \left( w \ln w + (1 - w) \ln(1 - w) \right), \tag{2.62}$$

which can be expressed either as a function of $U$ by making use of Eq. (2.60), or as a function of $\beta$ by making use of Eq. (2.61). This system exhibits the following behavior (see Figure 2.5). If all two-level systems are in the ground state, $w = 0$, then $\beta \to \infty$ (or $T \to 0$) as you would expect. Now, if $T$ is increased and becomes larger and larger and goes to infinity, $\beta$ is positive and becomes smaller and smaller and goes to zero, while $w$ goes to $1/2$. At very high temperature ($k_B T \gg E$) the states are thus equally populated ($w = 1 - w = 1/2$; of course, for $M$ states, one would get in this limit $w = 1/M$). All states with $0 < T < \infty$ could be prepared by putting the system in contact with an appropriate heat reservoir. But now, imagine you would prepare the system in a state with $U > NE/2$, i.e., $w > 0.5$, and then immediately isolate it. Looking at Eq. (2.61) makes clear that $w > 0.5$ has negative $\beta$. The larger $w$ (or $U$) gets, the larger the modulus is of the *negative* $\beta$. The extreme case would be, if you put all particles in the upper level, such that $U = NE$, or $w = 1$, which corresponds to $\beta \to -\infty$.

Although not often observed in practice (because most of real systems have energy spectra that are not bounded from above and have thus positive temperature - and the systems with bounded energy have contact with them), one can learn from this example an important fact on the nature of the temperature scale. Since positive and negative temperatures are naturally connected via $\beta = 0$, i.e., $T = \infty$, and not via $T = 0$, $\beta$ is the more convenient quantity to deal with. The two zero-temperature limits, $T \to 0^+$ and $T \to 0^-$ from the positive and negative sides, respectively, correspond to completely different states. At $T = 0$ a discontinuity occurs, while $T = \infty$ (or $\beta = 0$) corresponds to a continuous transition through $w = 1/2$. The statement of the third law, that absolute zero cannot be reached is thus not so astonishing at least for this system. If you prefer a measure for the temperature which algebraically increases if the system becomes hotter, you could use $-\beta$, which equals $-\infty$ at $T = 0^+$. The quantities $T$, $U$, $S$, and the heat capacity $C = T \, dS/dT$ (see Section 2.2.3) are illustrated in Figure 2.5 as functions of the parameter $-\beta E$.

Without going into details, two remarks are added just for interest (you may consult solid state textbooks for more details). In the case where the two energy levels correspond to the two opposite magnetic spin states (and $E$ is proportional to the magnetic field), the total magnetization is proportional to $w_1 - w_2 \propto \tanh(\beta E/2)$ (which you get from Eq. (2.61)) and describes an *ideal*

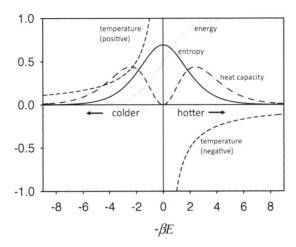

**Figure 2.5** Equilibrium states of the two-level system with positive and negative temperatures. The quantity $-\beta E$ (for constant $E$) is chosen as abscissa, because $-\beta$ gives a natural intuitive order of hotter as one moves to the right. Energy, temperature, entropy, and heat capacity are plotted in the natural units of $NE$, $E/k_B$, and $Nk_B$.

*paramagnet* for spin $1/2$. In fact, negative temperatures were experimentally realized in magnetic systems. Secondly, the maximum of the heat capacity as a function of $T$ (see Figure 2.5) is known as *Schottky anomaly*, and is sometimes used for describing the specific heat of glasses and amorphous materials.

Later we will discuss thermodynamic cycle processes, and negative temperatures could be included there, but we will not do it. If you are interested in that you may have a look at Ref. [LTAM80]. What you should keep in mind, is the formal association of negative temperatures with inverse populations, where states with higher energies are more probable than states with lower energies.

## 2.2  REVERSIBLE PROCESSES

The previous section was restricted to equilibrium steady states in isolated systems. The intensive quantities, like temperature and chemical potential, appeared naturally in the form of Lagrange multipliers. It was not yet necessary to mention the coupling of a system to an environment, like to a heat bath which sets the temperature, or to other reservoirs which set the chemical potential, the pressure etc.. Now we shall be concerned with *reversible processes* which refer to changes of states in the way that only equilibrium states are passed through. Because such state changes can only be performed when the system is coupled to an environment, described by reservoirs that act on the system, we must now first discuss system−environment couplings.

## 2.2.1  SYSTEM AND ENVIRONMENT

One of the first things a scientist learns at the beginning of his education is that a *system* is the part of the world which is under study, and its complement is the *environment*. You are free to put the system boundaries, and you may choose arbitrary properties for them (as long as they are physically consistent). In thermodynamics it is logical to define them with respect to the exchange of quantities like energy, particles, etc.. A thermodynamic system is said to be

**Isolated** if there is no exchange with the environment at all,
**Adiabatically closed** if there is no exchange of heat and matter,
**Closed** if there is no exchange of matter,
**Open** otherwise.

There exist two qualitatively different types of state quantities, namely, *extensive* variables (their magnitudes scale linearly with the system size in the limit of large size; examples are volume $V$, energy $U$, particle number $N$, mass $m$), and *intensive* variables (they are size independent; examples are the Lagrange multipliers temperature $T$, chemical potential $\mu$, and pressure $p$). It is often convenient to express extensive quantities in a system-size independent form, namely, in terms of *densities* (e.g., per volume, per mass, or per mole; examples are mass density $\rho = m/V$ and energy density $u = U/V$).

## 2.2.2  HEAT AND WORK

In order to discuss changes in the state of a thermodynamic system, the notions *heat* and *work* are introduced. Isolated systems in thermodynamic equilibrium are rather boring, since they are neither accessible for changes or thermodynamic processes, nor do they change their macroscopic state spontaneously. In the framework of our picture of energy levels, $E_i$, and occupation probabilities, $w_i$, an external, small action means to change these quantities by small amounts $dE_i$ and $dw_i$. How this is done in detail, is irrelevant at the moment - but it is obvious that the system cannot be isolated if it is possible to change its state from the outside. Such changes are illustrated in Figure 2.6. Part (ii) of the first law of thermodynamics explains how the energy can change. Denoting the energy by $U$, the change due to work by $\delta W$, and the change due to heat by $\delta Q$, the first law can be written as

$$dU = \delta Q + \delta W. \tag{2.63}$$

The notation $dU$ as a total differential refers to the fact that $U$ is a *state variable*. Heat and work are different. They do not describe the state, but

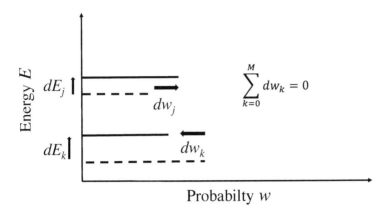

**Figure 2.6** A small change, $dU$, of the inner energy occurs due to small changes, $dE_i$ and $dw_i$, of the energy level values and the probabilities, respectively. The two contributions are associated with work and heat.

they characterize *how the state is changed* and are thus *process quantities*. Indeed, Eq. (2.1) implies a change of the average energy by

$$dU = N \sum_{i=1}^{M} E_i dw_i + N \sum_{i=1}^{M} w_i dE_i, \tag{2.64}$$

which reflects the first law (for simplicity we still consider $N$ fixed). Shifting the energy levels $E_i$ by $dE_i$ at constant occupation probabilities $w_i$ is just work:

$$\delta W = N \sum_{i=1}^{M} w_i dE_i. \tag{2.65}$$

The term related to the changes in the probability distributions $w_i$ by $dw_i$ at constant $E_i$ is then identified with heat,

$$\delta Q = N \sum_{i=1}^{M} E_i dw_i. \tag{2.66}$$

The $\delta$ indicates that the sum is in general not a total differential (which is denoted by $d$). From your basic math course, you know that for any total differential, $df$, a complete cycle $\gamma$ (where final state equals initial state) in a singly connected domain has the closed integral

$$\oint_\gamma df = 0 \tag{2.67}$$

(do not confuse this dummy $f$ with the distribution function from the previous section); $f$ is a *unique* function of the state, and thus also the final and initial values of $f$ are the same. This is not the case for work and heat. The work

$$\Delta W = \int_\gamma \delta W. \tag{2.68}$$

done after a finite change can thus be *dependent* on how the changes are done, i.e., it depends on the path $\gamma$ in the space of the parameters that are changed. The same holds, of course, for the heat

$$\Delta Q = \int_\gamma \delta Q. \tag{2.69}$$

A general relation for the work and heat integrals is, of course,

$$\oint_\gamma \delta Q = -\oint_\gamma \delta W, \tag{2.70}$$

for closed paths $\gamma$, which follows from Eq. (2.67) with $f = U$. The definitions of (2.65) and (2.66) are still general, i.e., independent of whether thermodynamic equilibrium holds or not. Nevertheless, we presume now that the changes are performed such that thermodynamic equilibrium holds (*reversible process*), i.e., the distributions satisfy $w_i = \exp(-\beta E_i)/Z$ in a quasi-stationary manner. Then, with Eqs. (2.6), (2.66), and $\sum dw_i = 0$ you get

$$dS = -Nk_B \sum_{i=1}^{M} \ln(w_i)\, dw_i = Nk_B \sum_{i=1}^{M} \beta E_i dw_i = \frac{\delta Q_{rev}}{T}, \tag{2.71}$$

where the subscript *rev* indicates the reversibility of the heat addition (in the following, we will add this subscript only when necessary for preventing confusion). In other words, if heat is added such that the state passes through thermodynamic equilibrium states (so-called *reversible* changes), it holds that $\delta Q = TdS$. The *reversible heat change divided by the temperature is the change of a state variable*, namely, $dS$. Equation (2.67) thus holds also for $df = \delta Q_{rev}/T$. This implies that if the state does not change reversibly, you can still calculate the entropy change $\Delta S$ between final and initial states by appropriately adding or subtracting heat (of course, provided they are both equilibrium states). But you must do it along a reversible path, even if the true process was irreversible. As a further consequence of the relation between heat and entropy, we can now write the first law in the form

$$dU = TdS + \delta W, \tag{2.72}$$

an equation which turns out to be very useful, because there is only the work part left which does not have the form of a total differential.

So let us now turn to the work $\Delta W$. According to Eq. (2.65), the microscopic picture is connected to changes in the energy levels. You certainly know a number of different work types connected to macroscopic variables. The general expression reads

$$\delta W = \sum_k Y_k dX_k, \tag{2.73}$$

where $X_k$ and $Y_k$ are, respectively, various extensive and conjugate intensive quantities. The relation between them and the energy levels $E_i$ can be very complex. The different types of work are often named according to the extensive quantity; here we list a few:

**Volume work** $X = V$ = volume, $Y = -p$ = pressure
**Magnetization work** $X = \vec{M}$ = magnetization, $Y = \vec{B}$ = magnetic flux density
**Electric charging work** $X = q$ = charge, $Y = \mathsf{U}$ = electric potential
**Electric polarization work** $X = \vec{P}$ = polarization, $Y = \vec{E}$ = electric field
**Chemical work** $X = N$ = particle number, $Y = \mu$ = chemical potential
**Acceleration work** $X = \vec{p}$ = total momentum, $Y = \vec{v}$ = center of mass velocity

The last example concerns mechanical work associated with the kinetic energy of the system considered as a whole body of mass. There are other work types like those related to the rotational energy of the system as a whole, or vibrational energy of macroscopic sub-parts of the system. They all are of the form (2.73), where $X$ and $Y$ are scalars, vectors, or tensors of higher ranks. Because the scalar quantity $\delta W$ cannot depend on how you chose your Cartesian coordinate system, it should remain invariant under rotations of your coordinate system. Therefore, scalars (which are not transformed under rotation) appear only in products with scalars, vectors with vectors (because their scalar product is obviously a scalar), and generally tensors with tensors of the same rank. This is rather general and will reappear below in Section 3.2 in the context of Curie's principle.

## 2.2.3 THERMODYNAMIC RESPONSE COEFFICIENTS

A large part of thermodynamics is concerned with the determination of thermodynamic equilibrium response coefficients, which relate small changes of a thermodynamic variable to small (reversible) changes of another variable. Mathematically these coefficients are nothing but partial derivatives of some functions, which you can immediately derive if you know the equilibrium state equations for your system. The main issue by performing a partial derivative in a space having more than a single dimension is, in which direction you differentiate, i.e., how you conduct the reversible process. There are various alternatives, like isobaric (constant pressure $p$), isothermal (constant $T$), adiabatic ($\delta Q = 0$), etc., changes. The corresponding response coefficients are then named accordingly, like the *isothermal compressibility* $(-V^{-1}(\partial V/\partial p)_T)$, the *isobaric expansion coefficient* $(V^{-1}(\partial V/\partial T)_p)$, the *isochoric coefficient of pressure* $(-p^{-1}(\partial p/\partial T)_V)$, or the *heat capacities*. We will only briefly discuss the latter for illustration and because it is relevant for heat storage. Other coefficients will be derived when needed, and you will find them also in standard textbooks.

We start with the *heat capacity*

$$C_X = \left(\frac{\delta Q}{\partial T}\right)_X = \left(T\frac{\partial S}{\partial T}\right)_X \tag{2.74}$$

of a system under the condition of zero work. The index $X$ indicates that all the $X_i$ are kept constant (no work) and the heat capacity corresponds to the heat change per temperature change. With $dU = \delta Q$ it holds that

$$C_X = \left(\frac{\partial U}{\partial T}\right)_X = N\frac{d}{dT}\frac{\sum E_i \exp(-\beta E_i)}{\sum \exp(-\beta E_i)} \tag{2.75}$$

where the $E_i$ are kept constant since no work is done. With $\beta = 1/k_B T$, the differentiation is elementary and yields

$$C_X = \frac{N}{k_B T^2}\frac{Z\sum E_i^2 e^{-\beta E_i} - (\sum E_i e^{-\beta E_i})^2}{Z^2} = Nk_B\frac{\langle E^2\rangle - \langle E\rangle^2}{(k_B T)^2} \tag{2.76}$$

where $\langle E^2\rangle$ is the mean squared energy. Note that $U = N\langle E\rangle$; the heat capacity $C_X$ is a measure for the energy variance.

If work is performed, the heat capacity attains a different value, because the heat is process dependent. As an example, we assume a system with work $\delta W = Y dX = -p dV$, and determine the heat capacity $C_p$ at constant pressure. Suppose the state functions $U(T,V)$ and $V(T,p)$ are known. With $\delta Q = dU + p dV$ it holds that

$$C_p = \left(\frac{\partial Q}{\partial T}\right)_p = \left(\frac{\partial U}{\partial T}\right)_p + p\left(\frac{\partial V}{\partial T}\right)_p \tag{2.77}$$

and with $(\partial U/\partial T)_p = (\partial U/\partial T)_V + (\partial U/\partial V)_T(\partial V/\partial T)_p$ and the definition of $C_V$ (for $X = V$) one obtains

$$C_p = C_V + \left(\left(\frac{\partial U}{\partial V}\right)_T + p\right)\left(\frac{\partial V}{\partial T}\right)_p. \tag{2.78}$$

One can express the second term on the right-hand side in different ways by using the first law and various relations between different derivatives of state functions. We do not need this here and will discuss below only a simple example, where the heat capacity ratio also becomes a microscopical physical meaning.

As a second case, we consider particle exchange, since energy conversion is often accompanied by it. In contrast to the previous example, $N$ is now a variable. The chemical potential $\mu$ as a Lagrange multiplier is associated with a particle number constraint for the entropy maximization. We neglect volume work for the moment and consider the entropy $S(U, N)$ as a function of the two variables, $U$ and $N$. The equation $dU = TdS + \mu dN$ implies

$$dS = \left(\frac{\partial S(U, N)}{\partial U}\right)_N dU + \left(\frac{\partial S(U, N)}{\partial N}\right)_U dN = \frac{dU}{T} - \frac{\mu}{T}dN. \qquad (2.79)$$

where Eq. (2.16) was used, and the relation

$$\left(\frac{\partial S(U, N)}{\partial N}\right)_U = -\frac{\mu}{T} \qquad (2.80)$$

will become important below for particle transport. In the exact same way as for the derivation of Eqs. (2.17) and (2.18) in Section 2.1.3, you can now discuss a system with subsystems at different chemical potentials, and show that equilibrium requires equality not only of $T$ but also of $\mu$. The same procedure can be repeated with other intensive variables $Y$.

By the way, Eq. (2.79) implies for bosons which have $\mu < 0$, that the addition of particles at constant $U$ leads to an entropy increase. This reflects the affinity of bosons for agglomeration, which already gives a hint of phenomena like Bose condensation or photon bunching, which we will not discuss further.

## 2.2.4  THERMODYNAMIC POTENTIALS

The entropy is at a maximum in isolated equilibrium systems. But what if a system, call it $\Sigma$, is not isolated? The general procedure to answer this question is to include the environment by describing it by equilibrium reservoirs (heat bath, particle reservoir, pressure reservoir, etc.,) and adding them to $\Sigma$ in order to obtain an isolated total system. If entropy maximization is applied to this total system, it turns out that the equilibrium state of the system $\Sigma$ can be described by optimizing certain thermodynamic potentials; by their definition, they will be at a minimum. Why will these potentials be important for us below? Because nonequilibrium states are characterized by currents which are driven by nonzero gradients of these potentials. But this comes later.

Thermodynamic equilibrium *baths* or *reservoirs* are defined as systems (we indicate their quantities here with a subscript 0), for which the first law can be written in the simple form $dU = T_0 dS_0$ (heat bath), $dU = -p_0 dV_0$ (pressure reservoir), $dU = \mu_0 dN_0$ (particle reservoir), etc., as is illustrated in Figure 2.7. Reservoirs are assumed to be so large that the intensive quantity can be considered constant. Let us illustrate the thermodynamic potential for the case of a heat bath. The result will turn out to be

$$F = -k_B T \ln(Z_N), \qquad (2.81)$$

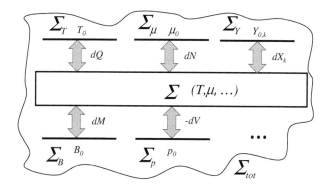

**Figure 2.7** The coupling of the system $\Sigma$ to its environment is described by the exchange of extensive quantities, $dX_k$, with separate reservoirs at fixed conjugate intensive quantities $Y_k$ (see Eq. (2.73) and the subsequent list). Specific examples are heat and particle reservoirs at temperature $T_0$ and chemical potential $\mu_0$, respectively, and various types of *work reservoirs*. For the total isolated system $\Sigma_{tot}$ (enclosed by the wavy curve), thermodynamic equilibrium is defined by maximum entropy.

which is called *free energy*. Equation (2.13) can then be written in the form

$$F = U - TS. \tag{2.82}$$

For reversible changes, where $dU = TdS + \delta W$, it holds that

$$dF = dU - TdS - SdT = \delta W - SdT. \tag{2.83}$$

Hence, the total differential $dF$ has now $T$ instead of $S$ as the independent thermal variable. By the way: $dF$ is a total differential since $F$ is defined as a function of the state variables $U$, $T$, and $S$, and therefore it is itself a state variable.

Now it will be shown that $F$ is the relevant potential when considering a system $\Sigma$ at constant temperature, i.e., in contact with a heat bath $\Sigma_T$. For this we make clear that $F$ is minimum in equilibrium. Additionally, we will derive that $F$ provides the maximum work that can be obtained from system $\Sigma$ at constant temperature.

The isolated total system is the combined system $\Sigma$ and $\Sigma_T$, and its entropy, $S_{tot} = S + S_0$, must be maximum. Since $dS_0 = \delta Q_0/T_0$, $\delta Q_0 = -dU$, and $T = T_0$, one obtains from the first equality in Eq. (2.83) with $dT = 0$

$$TdS_{tot} = -dU + TdS = -dF. \tag{2.84}$$

Hence, at constant $T$, maximization of the total entropy ($dS_{tot} = 0$) implies minimization of the free energy $F(T, X)$ of the system $\Sigma$ (i.e., $dF = 0$). This is why $F$ is called a *thermodynamic potential*.

| Name | Notation / Definition | Independent Variables |
|------|----------------------|----------------------|
| Energy | $U$ | $S, V, N, X$ |
| Entropy | $S$ | $U, V, N, X$ |
| Free Energy | $F = U - TS$ | $T, V, N, X$ |
| Enthalpy | $H = U + pV$ | $S, p, N, X$ |
| Free Enthalpy | $G = H - TS$ | $T, p, N, X$ |
| Grand Potential | $\Omega = F - \mu N$ | $T, V, \mu, X$ |
| General | $... - YX$ | $..., Y$ |

**Table 2.1**

**Different thermodynamic potentials with their independent variables.**

In order to understand, why $F$ is called also *free energy*, the situation can be interpreted as follows. We ask for the maximum work we can get from $\Sigma$. To answer this, we introduce an additional work reservoir, say $\Sigma_Y$, to which the work is provided. For $\Sigma_Y$ the first law reads $dU_Y = Y dX = -\delta W$, since this reservoir has by definition no entropy. The first law for the total system is

$$T dS_0 + dU + dU_Y = 0. \tag{2.85}$$

By making use of the second law, $dS_{tot} = dS_0 + dS \geq 0$, one finds for the work, $dU_Y$, delivered to $\Sigma_Y$

$$dU_Y = -dU - T dS_0 \leq -dU + T dS = -dF. \tag{2.86}$$

Recall that we are talking about $dF \leq 0$ and thus positive work provided by $\Sigma$, $dU_Y = -\delta W \geq 0$, to $\Sigma_Y$. Consequently, $dU_Y \leq| dF |$, i.e., *the free energy change is the maximum work that can be obtained from the system at constant temperature.* Equality holds in the reversible case, i.e., if $dS_{tot} = 0$. The total energy change, $dU$, can not be used. The part of energy which is *free* for use, i.e., which is available, is smaller by the amount of $TdS$, the heat transferred to $\Sigma_T$. We will come back to this topic soon when we introduce the terminology of *exergy* in Section 2.4.

You might have realized that the procedure to derive the thermodynamic potential can be generalized to each extensive variable $X_k$. If one fixes the intensity $Y_k$ by coupling $\Sigma$ to an associated reservoir $\Sigma_{Y_k}$, the thermodynamic potential is obtained by subtraction of $X_k Y_k$, such that the related energy differential becomes $Y_k dX_k - d(Y_k X_k) = -X_k dY_k$. This transformation between conjugate variables is called the *Legendre transformation*. Some important thermodynamic potentials are listed in Table 2.1. The way to use them is analogous to the free energy example and goes as follows. If you have a system where, for instance, $T$ and $p$ are given (as is often the case under real laboratory conditions), then the *free enthalpy* (or *Gibbs (free) energy*) $G = U + pV - TS$ is the appropriate thermodynamic potential that should be taken for determining equilibrium properties. Again, for the system $\Sigma$ equilibrium is not related to entropy maximization $dS = 0$ (which holds only for the isolated system), but to free enthalpy minimization $dG = 0$. This is important in chemistry, biology, and for discussing phase equilibria and the like. The procedure for other cases is analogous; we will not continue the discussion of thermodynamic potentials. At the moment we know enough for what we need later on.

Nevertheless, a few important general consequences of the thermodynamic potentials will now be added, which you should keep in mind. The first concerns the *Maxwell relations*. They are nothing but a manifestation of the symmetry of the second order derivatives of a real function $f(x_1, x_2, ...)$ of various variables $\{x_i\}_{i=1,2,...}$, i.e. $\partial^2 f/\partial x_j \partial x_k = \partial^2 f/\partial x_k \partial x_j$. For example, with $U(S, V)$ one gets from $T = (\partial U/\partial S)_V$ and $p = -(\partial U/\partial V)_S$, the relation

$$\left(\frac{\partial T}{\partial V}\right)_S = -\left(\frac{\partial p}{\partial S}\right)_V. \tag{2.87}$$

These relations connect different thermodynamic response coefficients to each other. Analogous relations follow from the other potentials in Table 2.1.

A second consequence is related to the homogeneity of the energy function $U(S, V, N_1, ...)$, where we allow different particle types with numbers $N_1, N_2, ..., N_n$, as is the normal case in chemistry. Since $U$ is a function of only extensive variables, which all scale linearly with the system size, a rescaling of the system by a factor $z$ leads to the functional form

$$U(zS, zV, zN_1, ...) = zU(S, V, N_1, ...). \tag{2.88}$$

If you differentiate this equation with respect to $z$, replace the resulting partial derivatives by their meaning as $T$, $p$, and $\mu$, and finally put $z = 1$, you get the *Euler equation*

$$U = TS - pV + \sum_{i=1}^{n} \mu_i N_i. \tag{2.89}$$

With this $U$, the *free enthalpy* $G = U + pV - TS$ takes the simple form

$$G = \sum_{i=1}^{n} \mu_i N_i. \tag{2.90}$$

which will be of use in Section 6.2.1. Another consequence of the Euler equation is obtained by taking the differential of Eq. (2.89), $dU = TdS + SdT - pdV - Vdp + \sum \mu_i dN_i + \sum N_i d\mu_i$, and equalizing it with $dU$ from the first law, $dU = TdS - pdV + \sum \mu_i dN_i$. This yields the *Gibbs–Duhem relation*

$$\sum_{i=1}^{n} N_i d\mu_i = Vdp - SdT, \tag{2.91}$$

which provides an additional relation between the *intensive variables*. They are thus not fully independent. Connected to this and more general is the Gibbs phase rule, which we will not derive. It provides the total number of free variables by $n_N - n_P + 2$, where $n_N$ is the number of different *species*, and $n_P$ the number of different phases (gaseous, liquid, ...). Of course, the trivial case is the single component gas with $n_N = n_P = 1$ and two free variables, e.g., $p$ and $T$. For a single component gas–liquid equilibrium, the number of phases increases to $n_P = 2$, such that only a single free variable is left (e.g., $T$, while $p = p(T)$ is then given by the vapor pressure relation). We will not need and discuss this topic further; it should be clear that the restriction of the number of free variables is due to fact that the different components are in thermodynamic equilibrium with each other. In *nonequilibrium* systems, the number of free variables can thus in general be larger. For instance, each species can have its own chemical potential and temperature.

## 2.2.5   EXAMPLES FOR REVERSIBLE PROCESSES

Let us now apply the previous formal theory to a few simple examples which illustrate the thermodynamic response coefficients and potentials for the ideal particle gas and the photon gas. The goal is not to *learn* about the specific final results, which you mostly know already anyhow. It is rather to understand the general procedure for the derivation of thermodynamic relations for the different processes, and to have some expressions at hand for use in later chapters.

### 2.2.5.1   The perfect gas

Important properties of the perfect gas are the heat capacities $C_V$ and $C_p$ (which are constant), and the thermodynamic equation of state, i.e., the relation between $p$, $T$, and $V$. First, Eqs. (2.46) and (2.75) lead to

$$C_V = \frac{3}{2} N k_B. \tag{2.92}$$

This result holds for classical monoatomic gases with three translational degrees of freedom. If more, say $f$, degrees of freedom are present (like rotations and vibrations of molecules), every degree contributes with $k_B T/2$. This *equipartition theorem* fails at very low temperatures when the classical physics breaks down - but this is not of interest at the moment. In the region where it holds, one has $C_V = f N k_B/2$. For instance, a diatomic molecule with two additional rotational degrees of freedom has $f = 5$. A generalization to, e.g., three rotation axes, vibrations, etc., is straightforward.

In order to determine the pressure, we use the first law, $dU = TdS - pdV$, and get

$$p = -\left(\frac{\partial U}{\partial V}\right)_S. \tag{2.93}$$

Since the derivative has to be done at constant entropy $S$, we make use of the Sackur–Tetrode relation (2.49), which tells us that at constant $S$ the expression $VT^{3/2} = K$ is constant and the caloric equation of state (2.46) can be written as

$$U = C_V T = \frac{3}{2} N k_B \left(\frac{K}{V}\right)^{2/3}. \tag{2.94}$$

The two previous equations yield the well-known thermodynamic equation of state of the ideal gas,

$$p = \frac{N k_B T}{V}. \tag{2.95}$$

Finally, since for the ideal gas $(\partial U/\partial V)_T = 0$, Eq. (2.78) leads to the heat capacity at constant pressure, or

$$C_p - C_V = N k_B. \tag{2.96}$$

The chemical potential $\mu$ can be calculated from Eq. (2.80) and the Sackur–Tetrode formula. The partial derivative must be done at constant $U$. Equation (2.46) suggests replacing $k_B T$, which is hidden in $\lambda_{th}$, by $2U/3N$ in Eq. (2.49) to obtain $S(U, N)$. The partial derivative with respect to $N$ at constant $U$ leads, with Eq. (2.80), to the chemical potential of the perfect gas

$$\mu = k_B T \ln \left(\frac{N \lambda_{th}^3}{V}\right). \tag{2.97}$$

If you like, you can get rid of $N/V$ and write this as a function of pressure and temperature with the help of Eq. (2.95); the chemical potential is then a function of the intensive quantities $p$ and $T$.

This example is important because the form $\mu = k_B T \ln(\rho_N/\rho_0)$ of the expression (2.97) for the chemical potential of a dilute particle density should be kept in mind. Here, $\rho_N \propto N/V$ is the density, and $\rho_0$ is a reference density. We will come back to this in Section 3.3.1.

We continue by looking at a few types of finite reversible changes (indicated by the $\Delta$). The most simple one is the *isochoric process* with constant volume, $dV = 0$. Then, $\Delta W = 0$ and the total change in inner energy becomes $(\Delta U)_V = (\Delta Q)_V$, where the subscript indicates that $V$ is constant. With Eq. (2.95), the relation between pressure and temperature changes is

$$(\Delta p)_V = \frac{N k_B}{V} (\Delta T)_V, \tag{2.98}$$

hence

$$(\Delta U)_V = (\Delta Q)_V = C_V (\Delta T)_V = \frac{1}{\kappa - 1} V (\Delta p)_V. \tag{2.99}$$

Here we used (2.96) and introduced the *adiabatic coefficient*

$$\kappa = \frac{C_p}{C_V}, \tag{2.100}$$

such that $Nk_B/C_V = \kappa - 1$. For a general number $f$ of degrees of freedom, it holds that $\kappa = 1 + 2/f$.

A second type of process refers to *isobaric changes* where the pressure $p$ is constant and the work is $-p(\Delta V)_p$. From Eq. (2.95) one concludes

$$(\Delta V)_p = \frac{Nk_B}{p}(\Delta T)_p, \tag{2.101}$$

and finds with Eq. (2.96)

$$(\Delta U)_p = C_p(\Delta T)_p - p(\Delta V)_p = C_V(\Delta T)_p = \frac{1}{\kappa - 1}p(\Delta V)_p, \tag{2.102}$$

with $-p(\Delta V)_p$ being the work.

A third type of important process refers to *adiabatic changes* where $\delta Q = 0$. This is the same as *isentropic changes* (constant entropy), as long as we have reversible processes, i.e., $TdS = \delta Q$ holds. It is obvious that $(\Delta U)_{ad} = (\Delta W)_{ad}$, but for calculating the work integral we have to determine first $p(V)$ along an adiabatic path. With $TdS = C_V dT + pdV$ and Eq. (2.95) one obtains with $dS = 0$ immediately $C_V dT/T = -Nk_B dV/V$, and after integration ($dT/T = d\ln T$, etc.):

$$TV^{\kappa - 1} = \text{constant.} \tag{2.103}$$

We can eliminate $T$ with the help of Eq. (2.95) and get

$$pV^{\kappa} = \text{constant,} \tag{2.104}$$

which leads to the work of the adiabatic process via the integral $-\int pdV$.

Of course, if the temperatures of initial and final states are known, one obtains the change in the inner energy directly from Eq. (2.46). The relations mentioned for the different processes can be helpful for calculating the efficiencies of machines which work along closed cycles that are composed of such processes, as will be discussed in Section 2.3.

We could now also derive, for a perfect gas, the various expressions for the thermodynamic potentials. We will do it when we need it, and mention here only the most straightforward example, which is the free energy. You can just use its definition $F = -k_B T \ln Z_N$, and then use Eq. (2.48). Also, the other potentials $H(S, p)$, $G(T, p)$, etc., follow, after some algebra, directly from their definitions in Table 2.1.

### 2.2.5.2 The photon gas

Equations (2.54) and (2.75) yield the heat capacity at constant volume,

$$C_V = 4aVT^3. \tag{2.105}$$

Furthermore, Eq. (2.59) tells us that constant entropy requires constant $VT^3 = K$, such that Eqs. (2.93) and (2.59) lead to the radiation pressure

$$p = -a\frac{d}{dV}\left(K^{4/3}V^{-1/3}\right) = \frac{a}{3}T^4 = \frac{U}{3V}. \tag{2.106}$$

(The factor $1/3$ can, by the way, also be understood if one knows that the pressure is in fact a tensor of the form $p_{jk} = p\delta_{jk}$, and that the energy density of the equilibrium radiation must be equal to the trace of the pressure tensor.) Because $p$ depends only on $T$ but not on $V$, one cannot change $T$ of the photon gas at constant $p$, and a definition of $C_p$ makes no sense. Other thermodynamic response coefficients (see Section 2.2.3) which require $p$ and $T$ as independent variables are also not defined for the photon gas; an example is the isothermal compressibility $-(\partial V/\partial p)_T$. The chemical potential is of course zero, $\mu = 0$, as has been discussed above.

Also, the thermodynamic potentials of the photon gas can easily be determined. For instance, from Eqs. (2.54) and (2.59) the free energy becomes

$$F = U - TS = -\frac{aVT^4}{3} = -pV = -\frac{U}{3} = -\frac{TS}{4}. \tag{2.107}$$

These few facts on the properties of the photon gas and the general concept of how to derive equilibrium relations should be sufficient for understanding what will be discussed in later chapters. More on the behavior of radiation will come later when we model radiative transport and solar energy conversion.

## 2.3  CYCLE PROCESSES AND EFFICIENCIES

After having understood thermodynamic equilibrium states and reversible changes, we are now ready to discuss the *efficiency* of reversible cycle processes. A thermodynamic cycle process can be seen as the representative of a periodically working engine. In an unusual way, we will start below not with the Carnot cycle but with the Otto cycle for a two-level system, because it helps to understand the fundamental reversible efficiency limit in terms of energy level $(E_j)$ shifts and probability $(w_j)$ re-distributions.

The general definition

$$\eta = \frac{\text{energy available for use}}{\text{energy input}} \tag{2.108}$$

of the energy efficiency indicates a certain ambiguity, because it refers to the specific use, or utility, and it must be specified closely for each single case what is meant by input and output energies. The net efficiency of a series of conversion processes with efficiencies $\eta_k$ is obtained by the product of their single efficiencies, $\eta = \Pi_k \eta_k$ which follows from the definition (2.108). If the output is of a mechanical, electrical, heat, etc., nature, the terms *mechanical efficiency, electrical efficiency, thermal efficiency*, etc., are common.

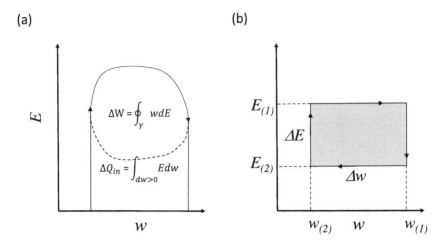

**Figure 2.8**  Two-level system with energy difference $E$ and excited state probability $w$ (see text). (a) Integrals associated with the added heat $\Delta Q_{in}$ and the total work $\Delta W$ in the $w$-$E$-plane associated with Eq. (2.109). (b) Otto cycle in the $w$-$E$-plane.

A *cycle process* for a thermodynamic system is defined by a closed path in its state space - as we discussed already in the context of the closed integrals of the form (2.67). The important thing is that the initial and final states of a system $\Sigma$ after a process cycle are identically the same, such that changes occur only in the environment. Reversibility means then that the total entropy of the environment does not change, which leads to the largest possible efficiency. For the moment, we focus on such reversible cycle processes; losses due to irreversibilities which decrease these maximum efficiencies are postponed to later chapters. A general energy efficiency needs of course not be related to a periodic process. For instance, for energy storage, one might sometimes be interested only in the discharge process, if the device is only charged once, and then one neglects the charging efficiency. The total (*round trip*) efficiency of the charge-discharge cycle is then the product of charging and discharging efficiencies - but this comes later in Section 6.

### 2.3.1  OTTO CYCLE EFFICIENCY OF A TWO-LEVEL SYSTEM

Maybe you have once derived in your thermodynamics lectures the efficiency of a cycle processes, like the Carnot process, for the perfect gas, by changing the macroscopic variables $T$, $V$, etc.. It is difficult to obtain from this a microscopic understanding of the final finding, that not all of the heat can be harvested in the form of work. In order to see better what happens during the cycle process in terms of energy-level shifts and changes in the probability distribution (remember Figure 2.6), it is helpful to discuss the Otto cycle for the two-level system (see Section 2.1.5.3). Since the efficiency is here size (or $N$) independent, we may consider specific quantities (per particle) by formally putting $N = 1$. The change in energy is then $dU = Edw + wdE$, with $\delta Q = Edw$ and $\delta W = wdE$. In this section the notation for $E$ and $w$ is *different* from the other sections. Here, $w$ is the probability of the higher energy level, $E = E_2$ is its energy, and $1 - w$ and $E_1 = 0$ are those for the lower level (ground state). In contrast to the notation in other sections, the subscripts (1) and (2) of $w$ and $E$ do not correspond here to different energy levels but to the values of $E$ and $w$ during the process, i.e., of the higher level during the cycle process described by a rectangular, closed curve in the $w - E$-plane (see Figure 2.8(b)). We can define the efficiency generally by the ratio of net work and heat input,

$$\eta = \frac{-\Delta W}{\Delta Q_{in}} = \frac{-\oint w\, dE}{\int_{dw>0} E\, dw} = \frac{\oint E\, dw}{\int_{dw>0} E dw}. \tag{2.109}$$

The second equality is due to the fact that $d(wE)$ is a total differential with vanishing integral after a cycle. These integrals are illustrated in Figure 2.8(a) by two areas. The four different process steps of an Otto cycle are illustrated in Figure 2.8(b). It consists of two adiabatic $(dw = 0)$ energy−level shifts by $\Delta E$ (for the ideal gas, this is the adiabatic volume change), and two heat exchanges $(\Delta w)$ at constant $E$ (for the ideal gas, this corresponds to isochoric changes). In the $w - E$-plane, this gives a rectangle parallel to the axes. The Otto cycle looks thus as follows in the statistical thermodynamics picture. At low occupation $w_{(2)}$ of the excited state with energy $E_{(2)}$, the energy level is increased by $\Delta E = E_{(1)} - E_{(2)}$, which requires the work input $w_{(2)}\Delta E$. The system is subsequently heated, which leads to an increase to $w_{(1)}$ in the occupation of the excited level at $E_{(1)}$. Afterward, the energy level is decreased at constant $w = w_{(1)}$, which provides a work output $w_{(1)}\Delta E$. Finally, the cycle is closed by returning to $w_{(2)}$ at constant $E = E_{(2)}$. The net work provided by the system is positive (or the work done at the system $\Delta W = W_1 - W_2 < 0$), and is given by the enclosed area of the process curve, hence

$$\mid \Delta W \mid = \Delta w \Delta E. \tag{2.110}$$

On the other hand, the heat input, given by the area below the $E_{(1)}$-line, is

$$\Delta Q_{in} = E_{(1)} \Delta w. \tag{2.111}$$

The efficiency turns out to depend on the ratio $E_{(2)}/E_{(1)}$. Now, denote by $\beta_{(j,k)} = \beta(w_{(j)}, E_{(k)})\ (\propto T_{(j,k)}^{-1})$ the value of $\beta = 1/k_B T$ at the corresponding corner of the rectangle. From Eq. (2.61), one concludes, $\beta_{(j,k)} = \ln(1/w_{(j)} - 1)/E_{(k)}$. Hence, the ratio of the energy shifts are $E_{(2)}/E_{(1)} = \beta_{(1,1)}/\beta_{(1,2)} = \beta_{(2,1)}/\beta_{(2,2)} = T_{(1,2)}/T_{(1,1)} = T_{(2,2)}/T_{(2,1)}$. The efficiency can thus be written as

$$\eta = 1 - \frac{E_{(2)}}{E_{(1)}} = 1 - \frac{T_{low}}{T_{high}}, \tag{2.112}$$

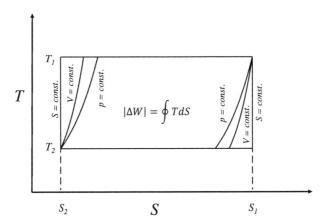

**Figure 2.9** Carnot, Stirling, and Ericsson cycle processes in the $S - T$-plane.

where $T_{low}/T_{high}$ is the ratio of lower and higher temperatures at the *same* *w-value* (the $T$-values are different for different $w$-values, but the ratio is constant). This is because Eq. (2.61) implies that $\beta E$ is constant for fixed $w$.

Now let us try to understand why the efficiency *must* be smaller than 1. Heat loss, $E_{(2)}\Delta w$, is necessary for depopulating the state $E_{(2)}$ from $w_{(1)}$ to $w_{(2)}$ (cooling) in order to be able to shift afterward the energy level up to $E_{(1)}$ with a *small* occupation probability (smaller work input). Only this enables the increase of the probability (by heating) such that $E$ can then be reduced from $E_{(1)}$ to $E_{(2)}$ with a *high* population (larger work output). Hence, there is no way around the cooling, and thus $\eta < 1$, except for $E_{(1)}/E_{(2)} \to 0$, i.e., $T_{low}/T_{high} \to 0$.

## 2.3.2 CARNOT EFFICIENCY

One recognizes in Eq. (2.112) already the typical form of the Carnot efficiency. Instead of deriving it for the ideal gas from adiabatic and isothermal processes as is usually done, we will use a simpler and more general approach. For this purpose it is important that Eq. (2.70) implies for a *reversible cycle*, that the net work obtained during the cycle can be calculated by the heat integral of $\delta Q = TdS$. For a reversible process one has

$$-\Delta W = \oint T \, dS. \tag{2.113}$$

A representation in the $S - T$ plane is thus natural. The most simple cycle process that can be illustrated in the $S - T$-plane has a rectangular area, i.e.,

(a)                                                    (b)

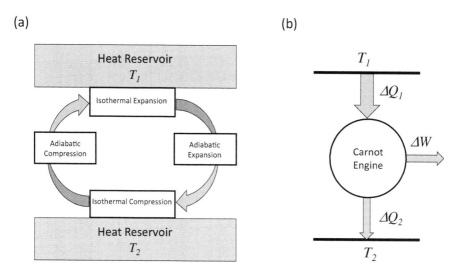

**Figure 2.10** (a) Carnot process cycle: a working fluid/gas is adiabatically and isothermally expanded/compressed by periodically cycling between two heat reservoirs at $T_1$ and $T_2$. (b): Carnot engine working between the reservoirs represented by in- and output energies.

it is bounded by isotherms ($dT = 0$) and adiabatic curves ($\delta Q = 0 = dS$), as shown in Figure 2.9. This is the Carnot cycle. There are other processes bounded by isotherms, like the *Stirling process*, which is additionally bounded by isochors ($dV = 0$) and the *Ericsson process*, which is bounded by isobars ($dp = 0$). Both are illustrated in Figure 2.9. All three have the same efficiency $\eta_0$.

Consider the Carnot process. The work is given by the area $\mid \Delta W \mid = (T_1 - T_2)(S_1 - S_2)$, and the heat input from the hot reservoir is $\Delta Q_1 = (S_1 - S_2)T_1$, the area under the upper isotherm. The heat flowing to the cold reservoir is $\Delta Q_2 = (S_1 - S_2)T_2$, the area under the lower isotherm. The Carnot efficiency is

$$\eta_0 = \frac{\mid \Delta W \mid}{\Delta Q_1} = 1 - \frac{T_2}{T_1}. \tag{2.114}$$

We derived it from a point of a view of state changes in the working fluid. A slightly different (but in fact identical) viewpoint is to look at the entropy changes in the reservoirs during a cycle. Consider the periodically working engine illustrated in Figure 2.10. The heat $\Delta Q_1$ (per cycle of the working fluid) is taken from the heat reservoir at constant temperature $T_1$ in order to produce the work $\Delta W$ while cooling the working gas to the temperature $T_2$ of the cold heat reservoir, to which the part $\Delta Q_2$ of the heat is added. The working fluid is then brought again to the higher temperature bath for heating up, and the cycle reiterates. Applying energy conservation (first law of thermodynamics) gives

$$\Delta Q_1 = \Delta W + \Delta Q_2 \tag{2.115}$$

such that the efficiency can be expressed in terms of the heat differences,

$$\eta = \Delta W/\Delta Q_1 = 1 - \Delta Q_2/\Delta Q_1 \tag{2.116}$$

This is still general, but now we suppose reversibility. The process is conducted through equilibrium states, i.e., the heat is added and removed from the reservoirs in a reversible manner. The entropy is then a total differential, which implies for an ideal cycle process that the total change in entropy (of the whole world) is zero:

$$\Delta S_1 + \Delta S_2 = -\frac{\Delta Q_1}{T_1} + \frac{\Delta Q_2}{T_2} = 0. \tag{2.117}$$

The system $\Sigma$ is in the same state after a cycle and has itself $\Delta S = 0$, because $S$ is a state variable. Replacement of the heat ratio in Eq. (2.116) with the temperature ratio from (2.117) gives the Carnot efficiency (2.114).

## 2.4   EXERGY AND ANERGY

Equation (2.114) means that the (maximum) *available* work $W$ which can be obtained from a heat, $Q$, is given by

$$W = \eta_0 Q \tag{2.118}$$

and depends on the temperatures of the involved heat reservoirs. Heat is thus energy with lower useful value as compared to the same amount of work. This motivated the thermodynamics engineers to introduce the terms *exergy* and *anergy* which divide up energy into *useful* and *useless* parts:

$$\text{Energy} \quad = \quad \text{Exergy} \quad + \quad \text{Anergy.} \tag{2.119}$$

Exergy is usually defined as the part of the energy, which can (for a given environment) be *transformed in all kinds of energies*. It is the maximum work, which can be obtained from a system embedded in a given environment. For instance, as we just learned, a heat bath at temperature $T_1$ with internal energy $Q$ (of course in the form of heat) has exergy $\eta_0 Q$ in an environment at temperature $T_2$. The anergy is the remainder $(1 - \eta_0)Q$. It cannot be transformed in exergy. For *heat*, Eq. (2.119) simply reads

$$Q = \eta_0 Q + (1 - \eta_0)Q. \tag{2.120}$$

You may thus claim, a Carnot engine is nothing but a device which separates the energy from a heat reservoir at $T_1$ into exergy and anergy by putting them, respectively, in a work reservoir and a heat reservoir at $T_2$, while keeping the total entropy (and of course energy) of the world constant. For more general cases one can still determine exergy and anergy: the exergy of inner energy, or the exergy of enthalpy, etc.. You should keep in mind that the exergy of a system depends on its environment.

We promised to discuss a connection between exergy and the thermodynamic potentials of Section 2.2.4, which are also related to a maximum work that can be obtained in a given situation. Let us derive the expression for the exergy of a closed system $\Sigma$ (cf. Figure 2.7), which is originally at temperature $T_1$ and pressure $p_1$ with energy $U_1$ and entropy $S_1$ and is brought to a state $p_2, T_2, U_2, S_2$ in equilibrium with the environment at $T_0$ and $p_0$ (such that $p_2 = p_0$ and $T_2 = T_0$). The energy change of $\Sigma$ is $\Delta U = U_2 - U_1$. In the reversible case (where one gets the maximum work), the entropy change of the heat bath is $\Delta S_0 = S_1 - S_2$, and the volume change of the pressure reservoir is $\Delta V_0 = V_1 - V_2$. The exergy, i.e., the work into the work reservoir, is denoted by $\Delta W_0$. Energy conservation of the total system $(\Delta W_0 + \Delta U + T_0 \Delta S_0 - p_0 \Delta V_0 = 0)$ implies for the exergy

$$\Delta W_0 = U_1 - U_2 + p_0(V_1 - V_2) - T_0(S_1 - S_2). \qquad (2.121)$$

The analogy with the free enthalpy $G$ is obvious (cf. Table 2.1). We learned in Section 2.2.4 that at constant temperature and pressure, the difference in $G$ is equal to the maximum possible work. The result (2.121) reflects this fact. The most simple example is the case of a pure heat reservoir, Eq. (2.118), which follows immediately from Eq. (2.121) with $V_1 = V_2$, if one replaces $T_2$ by $T_0$ and uses $U_1 - U_2 = Q$ and $S_1 - S_2 = Q/T_1$. In the last expression you need to take $T_1$ for getting the maximum available work, because any smaller value between $T_1$ and $T_2$ would give a smaller work.

Equation (2.121) shows that the second law is not the only reason why available energy can be smaller than the total energy. In addition to the entropic contribution $T_0 \Delta S$, there is also a volume-change contribution $p \Delta V$ (remember also what we said after Eq. (2.86) in Section 2.2.4). The work provided to the pressure reservoir, $-p_0 \Delta V_0$, is *not available* for use. This is particularly important in the gas flow where the expansion of the fluid has to be taken into account, which leads to the enthalpy $H = U + pV$ as the relevant quantity. The result (2.121) will be of use in Section 7.1.1.

The next chapter will deal with irreversible processes, which reduce the efficiency of a process even further below the unavoidable reductions discussed in this chapter. In engineering, sometimes the *exergetic efficiency*, $\eta_{ex}$ is used, which is defined by the ratio of output to input exergy of a process. For reversible processes, where the exergy remains conserved, $\eta_{ex} = 1$. The total efficiency can then be written as a product of the exergetic and reversible efficiencies, $\eta = \eta_0 \eta_{ex}$.

# 3 Linear Nonequilibrium Thermodynamics

Until now we considered equilibrium states and reversible processes that are sequences of equilibrium states. This is, however, not what happens in *real* processes. This chapter shows how to include *irreversibility* in the framework of linear nonequilibrium thermodynamics. Nonequilibrium states of an isolated system are (macroscopic) states whose probability distribution, $\{w_i\}$, deviates from the equilibrium distribution, $\{w_i^{(eq)}\}$. State changes, which go through nonequilibrium states are *nonequilibrium processes*. An example is relaxation to equilibrium. In practice, nonequilibrium processes often occur as an interplay of forces which drive the state away from equilibrium, and a response of the system which tries to restore equilibrium. This response looks like a counter-force which pulls the state towards the entropy maximum. We will look at two scenarios that are very important, namely, at *steady state currents*, and also briefly at *exponential relaxation*.

Unfortunately, there is no *simple general* theory for nonequilibrium processes beyond linear response (*far from equilibrium*). This should be obvious, if you imagine the huge number of possible microstates in a macroscopic system (in general, a nonequilibrium state may refer to any possible probability distribution $\{w_j\}$, and thus requires much more information than a few macrostate variables, like $T$, $\mu$, etc. that describe equilibrium states). Nevertheless, for sufficiently small deviations from equilibrium (i.e., near equilibrium) a simple general theory *does* exist, namely, when the deviations of all variables from their equilibrium values are so small that it is sufficient to restrict the consideration to the leading order of the deviation from equilibrium. In the following, we will therefore mainly focus on so-called *linear* nonequilibrium thermodynamics. Processes far from equilibrium, while very interesting, are still subject of modern research, but go beyond the scope of this primer on nonequilibrium physics. Let us now first understand what happens in an irreversible process.

## 3.1 IRREVERSIBLE PROCESSES

Reversible processes are theoretically infinitely slow, in order to ensure that the states remain in thermodynamic equilibrium. Every state of the quasi-stationary sequence of states maximizes then the entropy. *Slow* means here that changes occur on time-scales which are long as compared to the slowest intrinsic relaxation time. However, in reality changes occur in finite time, and the discipline which takes this into account is sometimes called *finite-time thermodynamics* in the literature [Bej96]. Of course, relaxation will always occur, but its relevance is a question of how fast the state changes (this may be different in the case of an instability, but this is excluded in near-equilibrium physics). The following example is simple and illustrative.

### 3.1.1  THE DROPPING STONE EXAMPLE

Consider a stone which is initially at temperature $T_1$, falls down from an altitude $h$, and hits the rigid ground (Figure 3.1 (a)). For simplicity, suppose that there is no energy transfer to the ground, i.e., the latter just reflects elastically the momentum of all the atoms which form the stone and which are supposed to be linked by forces (you may imagine springs). The highly ordered (initially potential, and then kinetic) energy of the atoms, will finally, when the stone is at rest, be statistically distributed (disordered) in vibrational degrees of freedom with zero mean velocity. This is illustrated in the $z$-$v$−plane shown in Figure 3.1 (b). Initially, the velocities of the atoms are distributed with zero mean in the dark rectangle of the altitude-velocity $(z, v)$-plane around $(h, 0)$. During the plunge the velocities of all the atoms follow a region around the solid arrow. Of course, the drawing is a bit misleading since in reality, the thermal velocities are much larger than the speed of fall - but it is at least a useful illustration. At impact, the distribution changes to the gray area near $z = 0$ having a larger velocity spread than initially at $T_0$, of course again with zero mean. This process is called *irreversible* since the reversed process is not running spontaneously (at least not within the age of the universe).

If the stone consisted of a single atom only ($N = 1$, without internal degree of freedom), it would just jump periodically up to altitude $h$ and fall down again. For a diatomic molecule ($N = 2$), the maximum altitude after the first bounce would be already less than $h$, because some energy will be transferred into rotational and vibrational energy (except in the unlikely case where the two atoms fall exactly horizontal to a flat ground). As $N$ increases, the jump height maximum will quickly decrease with time, and for sufficiently large $N$ the center of mass of the stone will be near the ground rather quickly. The exact behavior depends on the interaction strength between the atoms (there exist macroscopic, hard bodies which are elastically reflected from the hard ground and may jump up several times before coming to rest). There is a finite time when the macroscopic body is at rest and all the initially available potential energy is dissipated into irregular vibrational energy (i.e., heat), while no net energy at all was lost. The lost quantity is *exergy*! There is no artificial or hidden energy sink: this makes it clear that friction loss is nothing but a *redistribution* of the energy between ordered to disordered degrees of freedom.

Although the impact represents a highly irreversible nonequilibrium process, the produced heat and entropy can be calculated in a straightforward manner. The initial and final states are thermodynamic equilibrium states. The second law tells us then that for calculating the energy, we just have to add the corresponding heat in a *reversible* way, e.g., $\delta Q_{rev} = C\, dT$. With mass $M$ and heat capacity $C(T)$ of the stone, the initial gravitational energy $Mgh$

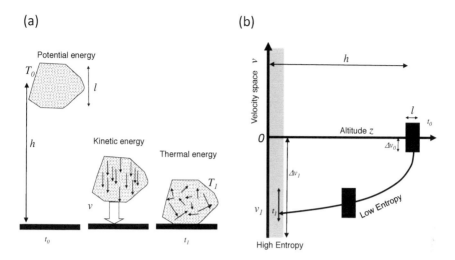

**Figure 3.1** Example of an irreversible process. (a) A stone is falling from an altitude $h$ down to the ground, increasing its temperature and entropy due to irreversible energy conversion from initially gravitational energy via kinetic energy to finally heat. (b) Cartoon of a microstate picture in the $z$-$v$-plane (representing phase space): at initial time $t_0$ the atoms are located around altitude $z = h$ and velocity $v = 0$ in a spatial region of size $l$ of the stone and velocity spread $\Delta v_0$ defined by the initial temperature $T_0$. The phase space volume (black, indicating a high density and low entropy) moves towards ground (collective motions), arrives at $t_1$ with velocity $v_1$ and, due to the impact, stops there while the velocity spread ($\Delta v_1$, gray region) increases (disordered motions, temperature-increase to $T_1$).

(with $g = 9.81$ ms$^{-2}$) must be equal to the final thermal energy,

$$Mgh = \int_{T_0}^{T_1} C(T)\, dT \approx C_0 \cdot (T_1 - T_0), \qquad (3.1)$$

where we assumed constant heat capacity, $C = C_0$, for the approximation. This yields the final temperature $T_1$ of the stone. With $TdS = C(T)dT$, the entropy increase becomes

$$\Delta S_{gen} = S_1 - S_0 = \int_{T_0}^{T_1} \frac{C(T)}{T}\, dT \approx C_0 \ln\left(1 + \frac{Mgh}{C_0 T_0}\right). \qquad (3.2)$$

For realistic values $Mgh \ll CT_0$ such that $\Delta S_{gen} \approx Mgh/T_0$: the generated entropy, $\Delta S_{gen}$, equals roughly the dissipated work (or the destroyed exergy) divided by the temperature ($T_1 \approx T_0$). Exact equality is obtained if the heated body can be seen as a heat bath, i.e., if $C_0$ is large. You will rediscover this result below.

### 3.1.2   IRREVERSIBLE CYCLE PROCESSES

Because equilibrium means maximum entropy, equilibration leads to an *increase of the entropy*, i.e., *entropy generation* or *entropy production*, $\Delta S_{gen} > 0$. This has an important implication for the efficiency of a cycle process. We will see in Chapter 5 that endoreversible thermodynamics provides an elegant framework to model efficiencies including irreversibility; however, in this section we look at it differently. We re-emphasize that the system, which represent the engine, is in the same state before and after the cycle process. But the environment can of course change its state: energy decreases in the hot reservoir and increases in the colder reservoir for a Carnot engine. The entropy influx from the hot bath into the system, $\Delta S_1 = \Delta Q_1/T_1$ and the outflux from the system into the colder bath, $\Delta S_2 = \Delta Q_2/T_2$, which are equal for the reversible Carnot process, differ now by the amount of the produced entropy, $\Delta S_{gen} > 0$, which must be removed in order for the engine to be in the initial state after the cycle. This entropy contribution is thus contained in the outflux:

$$\frac{\Delta Q_2}{T_2} = \frac{\Delta Q_1}{T_1} + \Delta S_{gen}. \tag{3.3}$$

Written for the system in terms of an integral along a general cycle (by definition, influx into a system contributes positively and outflux negatively):

$$\oint \frac{\delta Q}{T} = \frac{\Delta Q_1}{T_1} - \frac{\Delta Q_2}{T_2} = -\Delta S_{gen} < 0. \tag{3.4}$$

What does Eq. (3.3) mean for the *efficiency* of the cycle? Energy balance $| \Delta W | = \Delta Q_1 - \Delta Q_2$ still holds. The efficiency for irreversible processes is thus,

$$\eta = 1 - \frac{\Delta Q_2}{\Delta Q_1} = 1 - \frac{T_2}{T_1} - \frac{\Delta S_{gen} T_2}{\Delta Q_1} < \eta_0, \tag{3.5}$$

which is smaller than the reversible Carnot efficiency. This result can be formulated in terms of exergy: the heat $\Delta Q_1$ has an exergy $\eta_0 \Delta Q_1$. Now we have even less, $\eta \Delta Q_1$, i.e., we additionally lost $\Delta S_{gen} T_2$. The loss of exergy is obviously related to the entropy production

$$\text{exergy loss} = \Delta S_{gen} T_2. \tag{3.6}$$

We will see later that $\Delta S_{gen} T_2$ is in general the time-integral of the power dissipation.

### 3.1.3   MASTER EQUATIONS

From the discussion in Figure 3.1 (b), you may conclude that what really happens in an irreversible process is a time-evolution of the probability distribution, $\{w_j(t)\}_{j=1,...,M}$. Although a sophisticated discussion of the theory of stochastic processes and its applications in nonequilibrium physics goes far beyond the purpose of this book, this subsection touches on the issue because it is helpful for an intuitive understanding of irreversibility. You certainly divine what the theory formally looks like: ordinary or partial differential equations, which yield the time-evolution of the probability distribution. We will just demonstrate in the following how it works for the simplest case of linear nonequilibrium.

Suppose that equilibration can be described by ordinary differential equations for the $\{w_j\}$ of the form $dw_j/dt = F_j(\{w_k\})$, where the functions $F_j$ depend in general on all $w_k$ and other parameters of the system. In contrast to the very complicated general *strong nonequilibrium* (or *far from equilibrium*) case, the *weak nonequilibrium* (or *near-equilibrium*) cases can be described more easily, because they are by definition given by linearized equations. Schematically, these so-called *master equations* read

$$\frac{dw_j}{dt} = \sum_k (a_{jk}w_k - a_{kj}w_j). \tag{3.7}$$

The $\{a_{jk}\}_{j,k=1,\ldots,M}$ correspond to the transition rates from level $k$ to level $j$ (see Figure 3.2 (a)). Note that Eq. (3.7) is written in a form which satisfies probability normalization, i.e., $\sum dw_j/dt = \sum F_j = 0$. Consequently, one of the $w_j$ can be eliminated and one equation can be skipped. Other constraints, like conserved energy ($\sum E_j F_j = 0$ for isolated systems), etc., may further reduce the set of free variables; but we do not care here about such details. An important point, however, is that the $a_{jk}$ satisfy *detailed balance*

$$a_{jk}w_k^{(eq)} = a_{kj}w_j^{(eq)}. \tag{3.8}$$

In other words, in thermodynamic equilibrium, each parenthesis in Eq. (3.7) vanishes separately. If $a_{jk}$ is given, also $a_{kj} = a_{jk}\exp(-\beta(E_k - E_j))$ is known. Detailed balance means that, in the equilibrium state, the probability current (i.e., the rate $a_{jk}w_k^{(eq)}$) from level $k$ to level $j$ is balanced by the one from $j$ to $k$. For general nonequilibrium states, $\{w_j\}$, this so-called *micro-reversibility*, which is due to the time reversal symmetry of the fundamental microscopic equations, does not necessarily hold. For instance, stationary (circular) currents may exist which flow via more than two levels; an example is shown in Figure 3.2 (b).

Recall that thermodynamic equilibrium corresponds to the maximum of the entropy, provided certain constraints are satisfied. In the analogy of Figure 2.2 this means that the $\{w_k\}$ is restricted to a constraint sub-manifold. Assume now that by making use of all constraints, like normalization, energy conservation, and so on, the set of independent variables $\{w_k\}$ of $S$ is reduced by appropriate eliminations of some $w_j$, which become then functions of the remaining $w_k$. Obviously, in this sub-manifold in probability space, equilibrium refers to the vanishing of all (remaining) gradients $\partial \tilde{S}/\partial w_k = 0$, where $\tilde{S}$ is obtained via elimination of the dependent $w_k$. If some of these

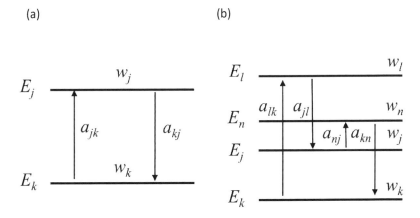

**Figure 3.2** (a) Transition rates $a_{jk}$ between different energy levels; in thermodynamic equilibrium Eq. (3.8) holds. (b) Example of a strong nonequilibrium case without detailed balance.

derivatives are non-zero, the probability distribution will relax to equilibrium with time, i.e., the rates of change, $dw_k/dt$ will in general be non-zero. Since weak nonequilibrium refers to confinement to the leading order deviations, the approximation with a linear relationship

$$\frac{dw_j}{dt} = \sum_k \gamma_{jk} \frac{\partial \tilde{S}}{\partial w_k} \qquad (3.9)$$

is sufficient, where $\gamma_{jk}$ is the proportionality matrix. The next step is to find the leading order, i.e. again linear, relation between the gradients $\partial \tilde{S}/\partial w_k$ and the deviations $\Delta w_j = w_j - w_j^{(eq)}$ from equilibrium. After combination, the result is a linear differential equation for the probabilities, which can model the equilibration behavior. By solving the (master) equations for the $w_j(t)$, you can determine the time-dependent macroscopic quantities which are linear combinations of the probabilities. This should be enough for a rough understanding of the microscopic background of irreversibility and equilibration, in order to perceive that entropy gradients can be seen as driving forces for equilibration. This will be important next.

## 3.2  GENERALIZED FORCES AND CURRENTS

If a thermodynamic quantity $X$ deviates slightly from its equilibrium value such that $dS/dX \neq 0$ is small, this quantity will feel a driving force in the direction of the entropy maximum, which is in leading order proportional to $dS/dX$. This force drives *currents* which tend to pull the state towards equilibrium. Remember that the entropy can be expressed as a function of

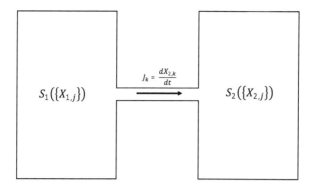

**Figure 3.3** An isolated system, consisting of two subsystems 1 and 2 with slightly different extensive quantities $X_{1,k}$ $(k = 0, 1, ..., K)$, and $X_{2,k}$ which are conserved $(X_{1,k} + X_{2,k} = X_k)$. The subsystems are weakly coupled in order to allow exchange of $X_k$ by currents $J_k$.

the extensive variables, $S(\{X_j\})$, $(k = 1, ..., K)$, and that its differential can be written in the form

$$dS = \frac{dU}{T} - \sum_{k=1}^{K} \frac{Y_k}{T} dX_k = -\sum_{k=0}^{K} \frac{Y_k}{T} dX_k, \tag{3.10}$$

where the $\{Y_k\}$ are the conjugate intensive variables. In order to shorten writing, $X_0 = U$ and $Y_0 = -1$ were introduced. Obviously, the partial derivatives of $S$ are

$$\frac{\partial S}{\partial X_j} = -\frac{Y_j}{T}, \tag{3.11}$$

where as usual all variables $X_k$ with $k \neq j$ are kept fixed for differentiation. In order to construct an isolated nonequilibrium system, we consider in Figure 3.3 two subsystems with slightly different values in $X_k$ (cf. also Figure 2.3). The goal is then to determine the associated thermodynamic forces, $\mathcal{F}_k$, and to relate them to the currents

$$J_k = \frac{dX_{2,k}}{dt} = -\frac{dX_{1,k}}{dt}, \tag{3.12}$$

of the extensive quantities from subsystem 1 into subsystem 2, by linear relations.

Let us come to the thermodynamic forces. We use the $X_{2,k}$ as the free variables, while the $X_{1,k} = X_k - X_{2,k}$ are determined, since the $X_k$ are conserved for the isolated total system. Consequently, with

$$S_{tot}(\{X_{2,k}\}) = S_1(\{X_k - X_{2,k}\}) + S_2(\{X_{2,k}\}) \tag{3.13}$$

and Eq. (3.11), the thermodynamic forces are

$$\mathcal{F}_k := \frac{\partial S_{tot}}{\partial X_{2,k}} = \left(\frac{Y_k}{T}\right)_1 - \left(\frac{Y_k}{T}\right)_2. \tag{3.14}$$

For example, if the $X_k$ for $k = 0, 1, 2$ denote energy $U$, volume $V$, and particle number $N$, respectively, one obtains with $-1$, $-p$, and $\mu$ for $Y_k$ the forces

$$\mathcal{F}_U = \frac{1}{T_2} - \frac{1}{T_1}, \tag{3.15}$$

$$\mathcal{F}_V = \frac{p_2}{T_2} - \frac{p_1}{T_1}, \tag{3.16}$$

$$\mathcal{F}_N = -\frac{\mu_2}{T_2} + \frac{\mu_1}{T_1}. \tag{3.17}$$

In equilibrium the $T$, $p$, $\mu$ are constant and the forces vanish. Otherwise, currents will flow. For example, if only $\mathcal{F}_U > 0$ then $J_U > 0$, if only $\mathcal{F}_V > 0$ then $J_V = dV_2/dt > 0$, etc.. This leads us to the second part: the linear relations between the forces and the currents,

$$J_l = \sum_{k=0}^{K} L_{lk} \mathcal{F}_k, \tag{3.18}$$

with the so-called *Onsager coefficients* $L_{lk}$. This is the second linearization after Eq. (3.14). There are three important properties of the Onsager matrix: *positivity*, the *Onsager–Casimir relations*, and the *Curie principle*. They are mentioned here without going into detail [LL13]. First we calculate the *entropy production rate*,

$$\frac{dS_{tot}}{dt} = \sum_{k=0}^{K} \left(\frac{\partial S_{tot}}{\partial X_{2,k}}\right) \frac{dX_{2,k}}{dt} = \sum_{k=0}^{K} J_k \mathcal{F}_k = \sum_{j,k=0}^{K} \mathcal{F}_j L_{jk} \mathcal{F}_k \geq 0, \tag{3.19}$$

which is the sum of the products of the thermodynamic forces and their conjugate currents. The non-negativity of the entropy production rate, which follows from the second law, implies that the $L_{lk}$-matrix is positive semi-definite [Kre81].

The *Onsager–Casimir reciprocity relations* are a consequence of the above-mentioned micro-reversibility, i.e., time-reversal symmetry of the microscopic equations (see Eq. (3.8)). They read

$$L_{jk}(B) = \epsilon_j \epsilon_k L_{kj}(-B), \tag{3.20}$$

where $\epsilon_l = \pm 1$ if the associated force ($\mathcal{F}_l$) is even/odd under time reversal, and $B$ stands for all those state variables that change sign under time reversal, like the magnetic field. The most prominent example of a case with magnetic field is the Hall effect in electric conduction, where the conduction matrix has off-diagonal components which change sign under time-reversal. In cases without magnetic fields and other time antisymmetric quantities, Eq. (3.20) just states symmetry

$$L_{jk} = L_{kj}. \tag{3.21}$$

The *principle of Curie* is related to spatial symmetries. In a spatially extended system, physical quantities can be classified according to the way they behave under spatial rotations. The different classes are represented by *scalars* (tensor of rank zero, like the temperature $T$), *vectors* (tensor of rank 1, like currents), and classes of *tensors* with higher ranks (like the stress tensor of rank 2). The Curie principle states then that, because $\dot{S}$ is a scalar, only currents and forces associated with tensors of the same rank are linked together, while $L_{jk} = 0$ if $j$ and $k$ refer to tensors with different ranks. For example, the gradient of a particle density is liked with a current density (both are vectors). It leads to a convenient simplification because it reduces the set of non-zero Onsager coefficients. A similar argument appeared at the end of Section 2.2.2. Despite the illustrative picture of *spatial* transport for forces and currents suggested by figure 3.3, the theory is of course general; obviously, you should now remember the last sentence of Section 2.1.3.

### 3.2.1 ENERGY AND PARTICLE CURRENTS

It is now time to illustrate the theory with a simple example, namely, energy and particle flow which are important for energy conversion applications later. In the case of temperature and chemical potential differences, the current-force relations can be written as

$$J_U = L_{UU}\mathcal{F}_U + L_{UN}\mathcal{F}_N \tag{3.22}$$

$$J_N = L_{NU}\mathcal{F}_U + L_{NN}\mathcal{F}_N \tag{3.23}$$

where we now use subscripts $U$ and $N$ for labeling the Onsager coefficients, and assume symmetry $L_{UN} = L_{NU}$. The eigenvalues of the Onsager matrix are

$$\lambda_{\pm} = \frac{Tr}{2} \pm \sqrt{\frac{Tr^2}{4} - det}, \tag{3.24}$$

with trace $Tr = L_{UU} + L_{NN}$ and determinant $det = L_{UU}L_{NN} - L_{UN}^2$. The positivity of the $\lambda_{\pm}$ requires

$$L_{UU} \geq 0 \quad , \quad L_{NN} \geq 0 \tag{3.25}$$

$$L_{UU}L_{NN} \geq L_{UN}^2. \tag{3.26}$$

The case of two closed subsystems where only heat but no particles can be exchanged ($L_{UN} = L_{NN} = 0$) is very simple. The Onsager matrix reduces then to the scalar $L_{UU}$, which is directly connected to the heat conductance. We shall consider this in the next subsection. Particle transport, on the other hand, can in general not be described solely by the scalar $L_{NN}$, because particles carry energy, such that Eq. (3.22) must not be disregarded. This example is helpful for understanding the meaning of the off-diagonal Onsager coefficients, here $L_{UN}$. Assume isothermal conditions $T_1 = T_2 = T$, which implies $\mathcal{F}_U = 0$. With Eq. (3.17) and $\Delta\mu = \mu_1 - \mu_2$, the particle current becomes

$$J_N = \frac{L_{NN}}{T}\Delta\mu. \tag{3.27}$$

With $\mathcal{F}_U = 0$, the energy current is

$$J_U = \frac{L_{UN}}{L_{NN}}J_N. \tag{3.28}$$

This tells us that the ratio $L_{UN}/L_{NN}$ has the meaning of the *energy which is transported by a particle* in the isothermal case (which is sometimes called the *transport energy*). The reciprocity relation, $L_{UN} = L_{NU}$, now has an interesting implication. If in the presence of a temperature difference, $\mathcal{F}_U \neq 0$, a steady state without particle current, $J_N = 0$, establishes, the chemical potential difference satisfies $\mathcal{F}_N/\mathcal{F}_U = -(L_{UN}/L_{NN})$. In words: the ratio of the two forces equals the transport energy.

### 3.2.2 RELAXATION

Until now we looked at stationary currents $J_k$ of the extensive quantities $X_k$. This is justified if the subsystems 1 or 2 are reservoirs, which have by definition equilibrium steady-states with constant intensive quantities $Y_{1,k}$ and $Y_{2,k}$. However, if they cannot be considered large, or if you wait long enough, a current $J_k$ will lead to a change of the intensive variables $Y_k$ of the subsystems, which no longer behave as reservoirs. Their time-dependence can be calculated by solving the linear differential equations defined by Eqs. (3.12) to (3.18). Let us show the principle for two thermally coupled heat reservoirs with an imbalance in temperatures, which leads to a heat or energy current. We write $\Delta U = U_1 - U_2$ and get

$$-\frac{d\Delta U}{2dt} = J_U = L_U\mathcal{F}_U = L_U(T_2^{-1} - T_1^{-1}) \approx \frac{L_U}{T^2}(T_1 - T_2), \tag{3.29}$$

for the linear approximation with $T \approx T_1$. One often writes the linear heat−flow law in the form

$$R_{th}J_U = T_1 - T_2, \tag{3.30}$$

which connects the temperature difference and the heat current $J_U$ from a hotter (1) to a colder (2) reservoirs separated by a thermal resistance $R_{th}$. Obviously, the heat conductance $R_{th}^{-1} = L_U/T^2$ is proportional to the Onsager coefficient $L_U$. Suppose that the heat baths have the same constant heat capacity $C_V$ and are in their separate equilibrium states at $T_1$ and $T_2$. It holds that

$$\frac{d\Delta U}{dt} = \frac{dU_1}{dt} - \frac{dU_2}{dt} = C_V\left(\frac{dT_1}{dt} - \frac{dT_2}{dt}\right). \tag{3.31}$$

Introducing the temperature difference $\Delta T = T_1 - T_2$, the combination of Eqs. (3.29)-(3.31) yields the relaxation equation

$$\frac{d\Delta T}{dt} = -\frac{\Delta T}{\tau}, \tag{3.32}$$

with energy relaxation time

$$\tau = \frac{C_V T^2}{2L_U} = \frac{R_{th} C_V}{2}. \tag{3.33}$$

It should be clear that similar relations like (3.30) hold for all other linear transport coefficients, like electric resistance, permeability, Peltier coefficient, and others you might know. The approach leading to the energy relaxation equation becomes analogous for other linear transport problems. For example, just replace heat by electric charge $Q$, temperature by electric potential $\mathsf{U}$, heat capacity by capacitance $C$, thermal resistance by electrical resistance $R$, and you will end up with the well-known RC-time $\tau = RC$. When you think it through, you might struggle with an obvious difference: the 2 in the denominator of Eq. (3.33) which is absent in the electrical $RC$-time. The reason lies in the long-range interaction of the electric charge: a potential shift $V_1$ at one electrode changes not only the charge by $Q_1$ at this electrode, but leads also to a negative charge of the same amount on the counter electrode. The total charge difference is then doubled - which is not happening in the thermal case above. Furthermore, in the multidimensional case of several forces and currents, the Onsager matrix must be considered, and one will find several relaxation times with the help of the methods of linear algebra. Because it makes the calculation more complex but does not bring us more insight, we continue now with the case of a single variable. The solution of Eq. (3.32) is

$$\Delta T = \Delta T_0 \exp(-t/\tau), \tag{3.34}$$

where $\Delta T_0 = \Delta U_0 / C_V$ is the initial temperature difference, and $\Delta U_0$ is the initial value of $\Delta U$. The time-dependent entropy production rate becomes, with $\mathcal{F}_U = \Delta T / T^2$,

$$\frac{dS}{dt} = J_U \mathcal{F}_U = L_U \mathcal{F}_U^2 = \frac{L_U}{T^4} \Delta T_0^2 \exp(-2t/\tau). \tag{3.35}$$

The total entropy change with Eq. (3.33) is thus

$$\Delta S_{gen} = \int_0^\infty dt \left( \frac{dS}{dt} \right) = \frac{C_V}{4} \left( \frac{\Delta T_0}{T} \right)^2. \tag{3.36}$$

Note that $\Delta S_{gen}$ does not depend on $L_U$ or $\tau$. The result (3.36) must be in accordance with the entropy difference between final and initial states. Within the linear approximation, we may assume the two reservoir temperatures initially at $T \pm \Delta T_0/2$, such that

$$\Delta S = C_V \int_{T+\frac{\Delta T_0}{2}}^{T} \frac{d\tilde{T}}{\tilde{T}} + C_V \int_{T-\frac{\Delta T_0}{2}}^{T} \frac{d\tilde{T}}{\tilde{T}}, \tag{3.37}$$

which gives

$$\Delta S = -C_V \left( \ln(1 + \frac{\Delta T_0}{2T}) + \ln(1 - \frac{\Delta T_0}{2T}) \right). \tag{3.38}$$

After expansion of the logarithms, with $\ln(1+x) \approx x - x^2/2 + ...$, in Eq. (3.38) to second order in $\Delta T_0/2T$, you will find for $\Delta S$ in leading order the result (3.36).

## 3.3  CURRENT DENSITIES

If the systems are spatially extended, current density vector fields $\vec{j}_k$ have to be considered. The associated current $J_k$ which flows through a specific area $A$ is obtained from

$$J_k = \int_A \vec{j}_k \cdot d\vec{A}, \tag{3.39}$$

where $d\vec{A}$ is the infinitesimal surface element, written as a vector which points in the direction of the surface normal. The current densities are often - but not always - related to the local gradients of intensive quantities, like temperature gradient, pressure gradient, and (electro-)chemical potential gradient. Nevertheless, current densities can also exist without a local gradient. Even our most important energy current density - heat radiation from the sun - belongs to this case. In fact, the photons propagate ballistically through vacuum space. So, let us in the next sections discuss briefly the two limit cases of *diffusive currents* and *ballistic currents*.

### 3.3.1  DIFFUSIVE CURRENTS AND LTE

In linear nonequilibrium thermodynamics, one can often make the additional assumption of *local thermodynamic equilibrium (LTE)*, which simply states that all infinitesimal volume elements can be considered as local equilibrium systems. In other words, two neighboring infinitesimal volume elements in space may be considered in the same way as the two subsystems - or reservoirs - depicted in Figure 3.3. The transport carriers (particles like atoms, molecules, electrons, ions, photons in opaque media, phonons) are then strongly interacting with the background medium which defines the local equilibrium reservoir. This scattering of the particle with the medium leads to *diffusive* behavior. Note that the term *diffusive* is here used in a general way: all quantities like momentum, energy, charge, etc., can propagate in a diffusive way; only when it concerns particle densities, will the term *diffusion* current be used. The picture of transport between the infinitesimal equilibrium reservoirs leads to the

(a)                                              (b)

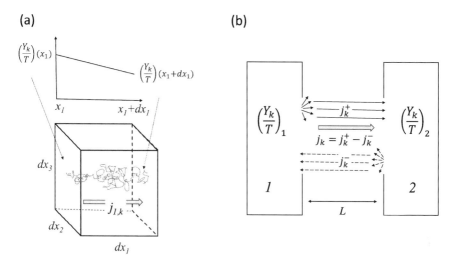

**Figure 3.4** (a) Diffusive current. The $x_1$-component of the current density $\vec{j}_k$ is proportional to the local gradient of $Y_k/T$; it can be defined in infinitesimal equilibrium volume elements, if the carriers behave in a diffusive manner and local equilibrium holds at all space points. (b) Ballistic current. If the carriers move without scattering from one reservoir to the other, the net current is the difference, $j^+ - j^-$, between the constant right- and left-moving currents ejected from the reservoirs. The conductance of a ballistic conductor is independent of its length $L$.

insight that the irreversibility, or more specifically: the entropy production rate, is also a density that is distributed in space. This will be discussed in the next chapter in more detail.

The assumption of LTE has two main consequences. First, the (thermal and caloric) *equations of states* are locally valid. All intensive quantities - like pressure, temperature, chemical potential etc. - become thus space dependent ($T = T(\vec{x})$, $p = p(\vec{x})$, $\mu = \mu(\vec{x})$, etc.). Extensive quantities, on the other hand, can be expressed as space dependent *densities*. For example, the particle number density at point $\vec{x}$ is given by

$$\rho_N(\vec{x}) = \lim_{\Delta V \to 0} \frac{\Delta N}{\Delta V}, \tag{3.40}$$

where $\Delta V$ is the small volume element centered at $\vec{x}$, and $\Delta N$ is the number of particles contained in $\Delta V$. Energy density $\rho_U$, entropy density $\rho_S$, etc., are defined in an analogous manner. The LTE equations of states can be written down in a straightforward manner. For example, the thermodynamic equation of state, $pV = Nk_BT$, and the caloric equation of state, $U = fNk_BT/2$, for the perfect gas of particles with $f$ degrees of freedom, as discussed in Section 2.2.5.1, become in terms of densities, respectively, $p = \rho_N k_B T$ and $\rho_U = f\rho_N k_B T/2$, or $p = 2\rho_U/f$.

Secondly, as the currents are related to differences of intensive variables, the current densities are proportional to spatial gradients of intensive variables. For example, consider two nearby space points $\vec{x}$ and $\vec{x} + d\vec{x}$, as shown in Figure 3.4 (a). It is sufficient to focus on a single space direction, say $x_1$, and a gradient of a single intensive variable, say $Y$ with conjugate extensive variable $X$. We can generalize the final result to higher dimensions and specific cases later. Therefore, suppose first $d\vec{x} = dx_1$, such that the thermodynamic forces Eqs. (3.14) can be written in the form

$$\mathcal{F}_k = -\frac{Y(x+dx)}{T(x+dx)} + \frac{Y(x)}{T(x)} = -\frac{\partial}{\partial x}\left(\frac{Y(x)}{T(x)}\right)dx. \tag{3.41}$$

The current, $J_i$, forced by $\mathcal{F}_k$ is given by $J_i = L_{ik}\mathcal{F}_k$ (see Eq. (3.18)). In this diffusive transport regime, $L_{ik}$, which has the meaning of a conductance or inverse resistance, is proportional to the area $A = dx_2\,dx_3$ and is inversely proportional to the distance $dx_1$ of the local reservoirs. You certainly know this geometry-scaling from the ohmic electrical conductance or from the heat conductance. We thus write $L_{ik} = \mathcal{L}_{ik}dx_2\,dx_3/dx_1$, with constant transport coefficient $\mathcal{L}_{ik}$ that characterizes the medium. The $L_{ik}$ are conduct*ances*, while the $\mathcal{L}_{ik}$ are conduct*ivities*. Because $J_i = j_i dx_2\,dx_3$, one has $j_i = \mathcal{L}_{ik}\mathcal{F}_k/dx_1$. Using Eq. (3.41), generalizing to three dimensions, and adding all relevant forces, one obtains

$$\vec{j}_i = -\sum_{k=0}^{K} \mathcal{L}_{ik}\vec{\nabla}\left(\frac{Y_k}{T}\right). \tag{3.42}$$

This is the local analog of Eq. (3.18). A simple example is a heat current density $\vec{j}_0$ due to a temperature gradient. With $Y_0 = -1$, one obtains

$$\vec{j}_0 = \mathcal{L}_{00}\vec{\nabla}\left(\frac{1}{T}\right) = -\lambda\vec{\nabla}T, \tag{3.43}$$

where we introduced the heat conductivity $\lambda = \mathcal{L}_{00}/T^2$.

You can easily derive further linear conduction laws analogous to Eq. (3.43). Because of its practical relevance, we will do this for a particle current with density and electrical potential gradient. Assume the particles have charge $q$, and the system has an electrical potential $\phi \neq 0$ different from the electrical potential $\phi = 0$ (ground) of the environment, the electric particle reservoir. The electrical work for changing the number of particles by $dN$ in the system is thus $\phi q\,dN$, which has to be added to the chemical work,

$$\delta W_{ec} = (\mu + q\phi)\,dN. \tag{3.44}$$

This leads to the replacement of the chemical potential by the *electrochemical potential*

$$\mu_{ec} = \mu + q\phi. \tag{3.45}$$

Suppose that the particle density and the electrical potential vary in space. Local thermodynamic equilibrium means that Eq. (2.97) holds at each space location. The local chemical potential is $\mu = k_B T \ln(\rho_N / \rho_N^{(0)})$, as was anticipated after Eq. (2.97). The expression (3.41) for the electrochemical force is thus

$$-\vec{\nabla}\left(\frac{\mu_{ec}}{T}\right) = -\vec{\nabla}\left(k_B \ln\left(\frac{\rho_N}{\rho_N^{(0)}}\right) + \frac{q\phi}{T}\right). \tag{3.46}$$

This is the generalized force acting on the particles. If we now apply Eq. (3.42) for obtaining the particle current density, $\vec{j}_{ec}$, it is clear that the Onsager coefficient $\mathcal{L}_{NN}$ is proportional to the particle density $\rho_N$, since every particle contributes to the current. For constant temperature $T$ and $\rho_N^{(0)}$, the particle current density can then be written as

$$\vec{j}_N = -b\rho_N \vec{\nabla}\left(q\phi + k_B T \ln \rho_N\right) = qb\rho_N \vec{E} - D\vec{\nabla}\rho_N, \tag{3.47}$$

where $\vec{E} = -\vec{\nabla}\phi$ is the electric field, and $qb$ is the mobility (usually denoted by the symbol $\mu$, which is reserved in this chapter for the chemical potential). I guess you have realized that the second expression on the right-hand side is nothing but Fick's law of diffusion. The diffusion constant $D$ obviously satisfies

$$D = bk_B T \tag{3.48}$$

which is the so-called *Einstein-relation*. It is a special case of the so-called *dissipation-fluctuation theorem*. We will not discuss it in detail, but you should know that it is again related to detailed balance. It states that, near thermodynamic equilibrium, fluctuations are linked to dissipation. The fluctuations are represented by the diffusion constant $D$ and the temperature $T$, as should be intuitively clear now from the meaning of temperature and diffusion. The link of $b$ to dissipation is also obvious from the origin of the Onsager coefficients. However, there is an even more direct connection to friction, as you probably remember from your courses on electric currents. The first expression on the right-hand side of Eq. (3.47) is just the drift current density $\rho_N \vec{v}$ with particle velocity $\vec{v}$. Considering Newton's equation for the particle with mass $m$, $md\vec{v}/dt = q\vec{E} - \vec{v}/b$, with electric and friction forces $q\vec{E}$ and $\vec{v}/b$, respectively, you get the velocity $\vec{v} = qb\vec{E}$ if the particle's inertia can be neglected. The mobility $qb$ characterizes friction and dissipation.

As it must be, Eq. (3.47) implies that in thermodynamic equilibrium, where $\vec{j}_N = 0$, the density is given by a Boltzmann distribution

$$\rho_N(\vec{x}) = \rho_N^{(0)} \exp\left(-\frac{q\phi(\vec{x})}{k_B T}\right). \tag{3.49}$$

Electric potential gradients and density gradients can exist in thermodynamic equilibrium - it is the *electrochemical* potential which is constant. As you will see in Section 7.1.2, such equilibrium electric fields are important in photovoltaic diodes, in order to separate the carriers generated by radiation in the semiconductor junction.

## 3.3.2   BALLISTIC CURRENTS AND NON-LTE

Local thermodynamic equilibrium (LTE) does not hold in *ballistic transport*. Transport is ballistic whenever the (quasi)-particles that carry the quantity under focus (energy, charge, mass, ...) are *not* scattered along their path from one reservoir to the other reservoir. An example for this type of non-LTE is *transport in rarefied gases*, where the density is so small that scattering is absent.

In a scattering event different properties can be exchanged in different ways. For example, scattering of an electron in a solid can be fully elastic, such that only the momentum is randomized and equilibrated but not the energy. In general there exist different characteristic lengths for momentum scattering and energy scattering. There is a continuous cross-over between diffusive and ballistic transport [Che05]. And scattering is not restricted to the kinetic degrees of freedom, for instance one may also have scattering energy transfer between spins, internal molecular states, etc.. In the following, we consider the most simple limit case of ideal ballistic transport where no scattering of the particles happens at all between two thermodynamic reservoirs. Besides photons in vacuum, there are many other applications of ballistic transport in modern technologies [Dat97, Che05], enabled due to the progressive miniaturization of devices, e.g., by nanotechnology. Examples are electrons in low-dimensional, so-called *mesoscopic*, conductors which can exhibit ballistic electric conduction. Also phonons, the quasi-particles associated with lattice vibrations in solids which transport heat, can behave ballisticly. Because of the convective character of ballistic transport, spatial gradients are not necessarily present locally despite the finite local current density, and the thermodynamic forces are related to the *finite differences* of intensive quantities between reservoirs. Figure 3.4 (b) illustrates this.

Let us calculate the heat current between two heat reservoirs for the case of heat radiation through a vacuum space (like the heat flow from the sun to earth). The non-interacting photons travel through free space with constant speed of light $c$, arrive on earth with an unchanged momentum distribution, and have thus approximately the temperature of the sun. We simplify the system by two blackbodies of temperature $T_1$ and $T_2 < T_1$ which are separated by a vacuum gap (in general this gap should not be too small, such that there is no interaction between the reservoirs by evanescent electro-magnetic waves [Che05]). Let us have a quick look back to Section 2.2.5.2 where we discussed the equilibrium photon gas, and determine the energy flux which leaves reservoir 1. Assume an open window in the reservoir, say perpendicular to the $x_3$-axis, where equilibrium radiation can escape. For the moment, we suppose a one-dimensional configuration $(x)$ space but three-dimensional wavenumber $(k)$ space. Only radiation with a positive momentum (or wave vector) component in the $x_3$-direction will escape from reservoir 1. Hence the photon distribution function in the vacuum gap is just given by Eq. (2.51) for $k_3 > 0$. Reservoir 1 does not contribute to the $k_3 < 0$ states in the vacuum.

Of course flux coming from reservoir 2 is doing that, but we can add this later and focus for the moment on the radiation from reservoir 1 alone. From Eq. (2.53) we conclude that the spectral energy density in wavenumber space is given by

$$u = \frac{2}{(2\pi)^3} \hbar c k f(\vec{k}),$$                     (3.50)

since integration over $d^3k$ yields the energy density $U/V$. The spectral energy *current* density is then $u\vec{c}$, where $\vec{c} = c\vec{k}/k$ is the velocity of the photons (that is of just the speed of light in the direction of flight). In order to get the total energy current density from reservoir 1 to 2, we need to integrate over the whole $k$-space with positive $k_3$. Since in spherical coordinates, $k_3 = k\cos(\theta)$ with $0 \le \theta \le \pi$, and the integration runs over the volume elements $d^3k = k^2 dk\, d\phi \sin(\theta)\, d\theta$ for $\cos(\theta) > 0$, we get in $x_3$-direction

$$
\begin{aligned}
j_U^+ &= \frac{2}{(2\pi)^3} \int_{\cos(\theta)>0} d^3k\, \hbar c^2 k \cos(\theta) f(\vec{k}) \\
&= \frac{cU}{V} \frac{1}{2} \int_0^1 \cos(\theta)\, d\cos(\theta) = \frac{acT_1^4}{4}.
\end{aligned}
$$                     (3.51)

Here, we used $d\cos(\theta) = -\sin(\theta)d\theta$, and the fact that only the $\theta$-integration differs between Eqs. (2.53) and (3.51). The factor $1/2$ in front of the second integral is to cancel out the factor 2 which comes from the $\theta$-integration over the whole solid angle when $U$ was calculated. The current density from reservoir 2 to 1 is obtained in the same way. The net heat current density between reservoirs 1 and 2 is

$$j_U = j_U^+ - j_U^- = \sigma_{SB}(T_1^4 - T_2^4),$$                     (3.52)

with the *Stefan−Boltzmann constant* $\sigma_{SB} = ac/4 = 5.67 \cdot 10^{-8}\ W/m^2K^4$ (for $a$, see Eq. (2.53)). You might object: "stop!" this current is not a linear function of the temperature difference, while claimed here to study *linear* nonequilibrium thermodynamics. But if you recall Eq. (2.54), you understand that $j_U$ in Eq. (3.52) is proportional to the difference in the energy densities, $U/V$, of the two reservoirs.

Let us conclude this section with the one-dimensional heat-transport example of ballistic phonons (one-dimensional electrical transport is postponed to the next chapter), e.g., in a linear chain molecule. Suppose, for simplicity, acoustic phonons which have the same linear dispersion relation ($\omega = ck$) and distribution function $f$ like photons (the result can thus be translated to photons). Of course, $c$ is now the acoustic wave speed and not the speed of light. Furthermore, we neglect different polarizations - you can easily multiply the final result with the corresponding degeneracy, i.e., the number of different polarizations, which is typically 3 for 2 transverse and 1 longitudinal polarizations. With phonon density of states $1/2\pi$ (see remark before Eq. (2.52)) the one-dimensional energy current becomes, with $dJ_U^+ = c\,\hbar\omega f dk/2\pi$,

$$
\begin{aligned}
J_U &= \frac{1}{2\pi} \int_0^\infty dk\, \hbar c^2 k (f(k, T+\Delta T) - f(k, T)) \\
&\approx \frac{\hbar c^2}{2\pi} \int_0^\infty dk\, k \frac{\partial f}{\partial T} \Delta T,
\end{aligned}
\tag{3.53}
$$

where we again subtracted right- from left-moving waves and approximated for linear response in $\Delta T$ at a given temperature $T$. After changing to the integration variable $s = \hbar ck/(k_B T)$, using Eq. (2.51), and performing integration by parts, this integral can be written as

$$
J_U = \frac{k_B^2 T \Delta T}{h} \int_0^\infty \frac{2s\,ds}{\exp(s) - 1} = \frac{\pi^2 k_B^2 T}{3h} \Delta T.
\tag{3.54}
$$

The interesting point here is the *universality*: this heat conductance $J_U/\Delta T$ is independent of the material, because the only material property involved, $c$, drops out. The heat conductance depends only on fundamental constants and the bath temperatures. This type of universality occurred already in the Stefan–Boltzmann law for radiation (because the speed of light is also a fundamental constant), and can even be generalized to arbitrary dimension and temperature differences (see, e.g., Ref. [DV08]). It is based on the ballistic character of the transport and the equilibrium character of the reservoirs. Nevertheless, non-ideal reservoir surfaces can influence the transport behavior by their properties.

The two above examples concerned bosons. We will see in the next chapter, that for one-dimensional *electrical* conduction (where the carriers are fermions at the Fermi level), the universality is even stronger in the sense that the electrical conductance is quantized in units of $e^2/h$, i.e., it depends *only* on fundamental constants, but not on the temperature.

# 4 The Entropy Production Rate

Two aspects make the entropy production rate, $\dot{S}$, highly relevant. First, $\dot{S}$ is *the* characteristic for equilibration and nonequilibrium processes. Secondly, $\dot{S}$ corresponds to the extinction rate of exergy, i.e., it decreases the efficiency of technical devices.

The first aspect might mislead to the tempting (but wrong) inference, that a general principle for *nonequilibrium processes* based on $\dot{S}$ exists, similar to the entropy maximization principle which holds for equilibrium states. After all, we have seen that derivatives of the entropy act as equilibration forces. The relevance of entropy for the evolution of natural processes was emphasized by Sommerfeld [Som96] who cited from a short *Nature* article by R. Emden: "In the huge manufactory of natural processes, the principle of entropy occupies the position of the manager, for it dictates the manner and method of the whole business, whilst the principle of energy merely does the bookkeeping, balancing credits and debits." It is the entropy that acts as the chief executive officer, while energy takes the role of the chief financial officer! Despite several attempts to find a *general* law for stationary nonequilibrium processes based on the entropy production rate, there is common agreement that *no* such generally valid law exists (for reviews, see Refs. [MS06, Ich94]). Applications of entropy maximization to nonequilibrium systems (like in so-called *extended thermodynamics*), and of entropy production optimization to systems beyond linear nonequilibrium cannot be justified in a strict manner, although they may sometimes provide useful trends for the behavior of nonequilibrium systems [MR98, DLNRL14]. Nevertheless, sufficiently close to equilibrium, and if the linear relations between forces and currents and the Onsager−Casimir symmetry relations (3.21) hold, it *is* possible to formulate an *entropy production optimization principle* for steady state currents. The trivial reason is that $\dot{S}$ (cf. Eq. (3.19)) is then a positive (semi-) definite bilinear form of thermodynamic forces $X_l$ and currents $J_l$.

Prior to a deeper discussion of the entropy production rate, let us comment briefly on the second aspect, namely, the equivalence of entropy production and loss of exergy. Since exergy loss is unwanted, it is consequent to use *entropy generation minimization* as a *design principle* for energy conversion and storage devices, as was discussed by Bejan and others [BTM96]. The fact that entropy production is unavoidable, combined with other design constraints, sometimes leads to a well-defined trade-off of irreversible effects, and thus to a minimum of $\dot{S}$ as a function of the controllable design parameters. The entropy generation minimization design principle, which is a $\dot{S}$-optimization principle in a completely different context than what is the topic of Section

4.3, will be postponed to Section 8.2 in the last chapter.

## 4.1  TOTAL ENTROPY PRODUCTION RATE

### 4.1.1  $\dot{S}$ IN STATISTICAL THERMODYNAMICS

Remember that the entropy Eq. (2.6) is a purely statistical quantity, i.e., it can be written as a function of the probability distribution only. This expression is also defined for nonequilibrium states, and one may ask how it changes with time during equilibration. In an isolated system, $\dot{S} = dS/dt$ is the (irreversible) entropy production rate. Applying the chain rule to Eq. (2.6) and using normalization, $\sum_j \dot{w}_j = 0$, one obtains

$$\frac{dS}{dt} = -Nk_B \sum_{j=1}^{M} \ln(w_j)\dot{w}_j, \tag{4.1}$$

which indicates that $\dot{S}$ is a weighted sum of the $\dot{w}_j$. If you plug $w_j = w_j^{(eq)} + \Delta w_j$ into Eq. (4.1), and expand the logarithm to linear order in $\Delta w_j / w_j^{(eq)} \ll 1$, you will find with the help of $w_j^{(eq)} = \exp(-\beta E_j)/Z_N$, normalization $\sum \Delta \dot{w}_j = 0$, and Eq. (2.66)

$$\dot{S} = \frac{N \sum_{j=1}^{M} E_j \dot{w}_j}{T} - Nk_B \sum_j \frac{\Delta w_j \Delta \dot{w}_j}{w_j^{(eq)}} = -Nk_B \sum_{j=1}^{M} \frac{\Delta w_j \Delta \dot{w}_j}{w_j^{(eq)}}, \tag{4.2}$$

where the first term is just the entropy rate $\dot{Q}/T$ associated to heat exchange with the environment, which vanishes because the system is isolated. The second term is the (irreversible) entropy production rate. During *equilibration* the $(\Delta w_j)^2$ decrease, thus it is clear that $\Delta w_j \Delta \dot{w}_j < 0$ and $\dot{S} > 0$. The total $dS/dt$ in an non-isolated system can of course have either sign, because the additional contributions, like $\dot{Q}/T$ describing heat exchange, can be positive (heat inflow) or negative (heat outflow). The same holds if particle exchange is involved, because the particles also carry entropy. However, for simplicity we focus now on the isolated system. It is worthwile to reiterate the characteristic structure of Eq. (4.2), which is again visible: $\dot{S}$ contains products of forces ($\propto \Delta w_j$) and currents ($\propto \Delta \dot{w}_j$). Of course, to relate the forces $\{\mathcal{F}_k\}$ and the currents $\{J_k\}$ to the $\{w_j\}$ and $\{\dot{w}_j\}$ is in general cumbersome - but due to the formal similarity and the fact, that macroscopic variables are usually linear combinations of the $w_j$, the analogy should be clear.

With Eq. (4.2), the total entropy production, $\Delta S$, becomes

$$\Delta S = \int_0^\infty \dot{S}\, dt = Nk_B \sum_j w_j^{(eq)} \frac{1}{2} \left( \frac{\Delta w_j^{(0)}}{w_j^{(eq)}} \right)^2, \tag{4.3}$$

where $\Delta w_j^{(0)}$ are the initial values of the deviations from equilibrium, and $\Delta w_j(t \to \infty) \to 0$. Equation (4.3) tells us that, in leading order, the entropy production is given by the averaged relative deviations from the equilibrium distribution. This result provides a microscopic, third way to derive Eq. (3.36). To show this, we need the expression

$$\frac{dw_j^{(eq)}}{dT} = \frac{E_j e^{-\beta E_j}}{k_B T^2 Z} - \frac{e^{-\beta E_j} \sum_i E_i e^{-\beta E_i}}{k_B T^2 Z^2} = \frac{w_j^{(eq)}}{k_B T^2}(E_j - \langle E \rangle). \qquad (4.4)$$

Note that $C_X = C_V$ in Eq. (3.36) was the specific heat of the two equal subsystems at initial temperatures $T + \Delta T_0/2$ and $T - \Delta T_0/2$. If the total system has size $N$, the subsystems have size $N/2$, such that Eq. (2.76) for their specific heats at zero work can then be written with the help of Eq. (4.4) as

$$C_X = \frac{N}{2k_B T^2}(\langle E^2 \rangle - \langle E \rangle^2) = \frac{N k_B T^2}{2} \sum_j w_j^{(eq)} \left( \frac{dw_j^{(eq)}/dT}{w_j^{(eq)}} \right)^2. \qquad (4.5)$$

On the other hand, we may replace in Eq. (4.3) the general deviations $\Delta w_j^{(0)}$ by the specific ones associated to the temperature shifts, $\Delta w_j^{(0)} = (dw_j^{(eq)}/dT) \cdot \Delta T_0/2$. If you now use the expression obtained with Eq. (4.5), you will immediately find Eq. (3.36) for Eq. (4.3).

### 4.1.2 $\dot{S}$ IN PHENOMENOLOGICAL THERMODYNAMICS

In many cases, $\dot{S}$ can be written down directly from simple phenomenological arguments. Consider an isolated system which contains (mechanical, electrical etc.,) exergy which is dissipated by some mechanism into heat at a rate $P = -\dot{W}$. Of course, the total energy remains constant, $\dot{U} = \dot{Q} - P = 0$. The entropy increases at a rate $\dot{S}$. Consider first constant temperature $T$. We know that $dS = \delta Q/T$, where $\delta Q$ is a *reversible* heat change (since you know that the true process is irreversible while the heat used for calculating the entropy must be reversibly added, we skip the subscript *rev*). We assume thus a reversible heat change with power $P$, i.e., $\delta Q = P\,dt$, hence

$$\dot{S} = \frac{P}{T}. \qquad (4.6)$$

This was rather simplistic - but wait a moment: we assumed constant temperature $T$, whereas if exergy is converted into heat, the temperature locally increases. To correct this, we need $\dot{S}$ for heat conduction. We know the result from the previous chapter: one considers two heat reservoirs as shown in Figure 4.1 (a) at temperatures $T_1 \geq T_2$, which are connected by a heat conductor

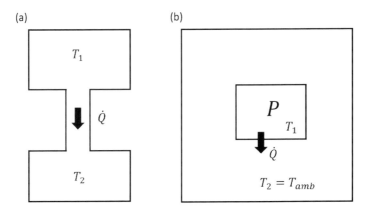

**Figure 4.1** (a) Heat conduction (heat rate $\dot{Q}$) between two reservoirs at temperatures $T_1$ and $T_2$. (b) Heat production by energy dissipation $P$ in a subsystem of temperature $T_1$ embedded in a large system at ambient temperature $T_2$.

that transfers a heat per time $\dot{Q}$. The heat change rates of the two reservoirs are

$$\dot{Q}_1 = -\dot{Q} = -\dot{Q}_2. \tag{4.7}$$

where $\dot{Q}_{1,2}$ are the heat currents into the reservoirs. The total entropy production rate $\dot{S}$ is the sum of the two entropy production rates $\dot{S}_k = \dot{Q}_k/T_k$ ($k = 1, 2$):

$$\dot{S} = \dot{Q}\left(\frac{1}{T_2} - \frac{1}{T_1}\right). \tag{4.8}$$

As we know from Eq. (3.19), the entropy production rate is the product of a generalized *current* $\dot{Q}$ and a *force* $\Delta(T^{-1})$.

Let us now come back to Eq. (4.6). The temperature $T$ is in general not constant, particularly not in the region of heat dissipation. Suppose the power dissipation $P$ heats up its nearby environment to a temperature $T_1$. From there, the heat flows into the far environment which has lower temperature, $T_2$ (see Figure 4.1 (b)). The whole system is still isolated, and the heat currents are quasi-stationary, i.e., in flow-equilibrium during the time of interest, and we consider the subsystems 1 and 2 as equilibrium reservoirs (of course, they are so large that they are not depleted during the time considered). The entropy rate of change of the bath at $T_1$ is now obviously $(P - \dot{Q})/T_1$. Because at steady state, $P = \dot{Q}$, the same argument as in the previous example gives for the total entropy production rate:

$$\dot{S} = \frac{P - \dot{Q}}{T_1} + \frac{\dot{Q}}{T_2} = \frac{P}{T_2}. \tag{4.9}$$

You can iterate this procedure. If the system under consideration is embedded in even a larger system with again another, lower temperature, $T_0$, into which the heat can flow, one has to replace $T_2$ by $T_0$ in Eq. (4.9). So, Eq. (4.9) holds for arbitrary $T$-distribution. It had to be expected and it is a rather important result: the *entropy production rate times the ambient temperature equals the heat production*, $P = T_0 \dot{S}$.

Let us close the circle via the exergy. In Eq. (3.6) we have seen that entropy production means exergy loss. $\dot{S}$ can be interpreted as *exergy destruction rate* and used as a quantification of energy dissipation. The exergy (2.121) of a closed system was derived for reversible processes by taking into account that the entropy change of the heat bath, $\Delta S_0$, was equal to the difference of the entropies of initial and final state of the system, or $\Delta S_0 = S_1 - S_2$. In the presence of an internal irreversibility in the system, the net change in entropy is positive, $\Delta S_0 = S_1 - S_2 + \Delta S_{irr}$. Eliminating $\Delta S_0$ in the original energy balance, the right-hand side of Eq. (2.121) obtains an additional part $\Delta W_{irr} = -T_0 \Delta S_{irr}$. Changing from differences ($\Delta W_{irr}$ and $\Delta S_{irr}$) to rates ($\dot{W}_{irr}$ and $\dot{S}$, where we skipped the subscript $irr$ at $\dot{S}$ according to our previous notation) this can be expressed as

$$\dot{W}_{irr} = -T_0 \dot{S}, \tag{4.10}$$

which is a form of the *Gouy–Stodola theorem* [Bej88]. The entropy production rate times the ambient temperature equals the *exergy destruction rate*. I guess that you certainly expected equality of exergy destruction and power dissipation rates, $\dot{W}_{irr} = -P$.

## 4.2 LOCAL ENTROPY PRODUCTION RATE

Let us now investigate the *local* entropy production rate in spatially extended systems. In line with Section 3.3, we separate between diffusive and ballistic transport.

### 4.2.1 $\dot{S}$ IN DIFFUSIVE TRANSPORT

It is now time to continue the discussion of Section 3.3.1 on entropy production rate *densities* in LTE systems. An important point is that the local volume elements are not isolated but open. In a volume $V$ that exchanges heat, work, particles, etc., with its environment, a finite change of entropy $dS/dt$ can occur without irreversible processes, namely, if there is a net difference between in- and outflux of entropy. In case of heat exchange, we mentioned already the term $\dot{Q}/T$ due to entropy flow in Eq. (4.2). In an open system, the entropy of a volume can change also due to flows of other extensive quantities, like particle flows, which carry their own entropy with them. Consequently, the entropy production rate, $\dot{S}_V$, in volume $V$ is now related to the total local entropy change rate $dS_V/dt$ in $V$ and to the entropy current density $\vec{q}_S$ by

$$\frac{dS_V}{dt} + \iint_{\partial V} \vec{q}_S \cdot d\vec{A} = \dot{S}_V, \tag{4.11}$$

where the surface integral extends over the surface $\partial V$ of the volume $V$. The surface-normal, $d\vec{A}$, of $\partial V$ points by definition outwards from the volume, while the fluxes are defined positive for inflow; hence in case of a net inflow, the surface integral is negative, and thus the corresponding entropy change is positive. If one moves now towards a description in terms of densities, as we started already in Section 3.3.1, Eq. (4.11) is applied to infinitesimal volume elements $\Delta V$.

Truly conserved quantities like mass, momentum, electric charge, and energy are governed by balance equations which are exact continuity equations for the associated local volume densities. Since entropy can be produced or reduced (at least locally), the associated balance equation (of which the global form is given by Eq. (4.11)), will contain source and sink terms. You might argue that the equations for conserved quantities are often also written with source and sink terms, like the power density in the energy balance equation, or generation-recombination terms in the ion-drift equations for electrolytes, or in the semiconductor equations for electrons and holes. This is, however, due to an artificial decomposition of the whole world in subsystems, and restricting the balance equation to one of those. For example, the decomposition of the energy of a fluid into heat and average kinetic energy leads to a power source, namely, the local dissipation of kinetic energy by friction, in the heat balance equation. The same can appear for momentum exchange of a fluid with a host medium, or particle generation by decomposition of a composite molecule. Nevertheless, conservation quantities are always *exactly* conserved in the *total isolated system* - while for the entropy, a true source can exist even in an isolated system.

It is instructive, to derive the local balance equation for the entropy from the balance equations for the truly conserved quantities. We will do it for the restricted case where only the volume densities $n(\mathbf{x}, t)$ and $u(\mathbf{x}, t)$ for particle number and energy are relevant. The plan is to identify the entropy current density, $\vec{j}_s$, and the local entropy production rate, $\dot{s}$, and to re-derive the finding $\dot{S} = P/T_{amb}$ made after Eq. (4.9). Volume work is neglected for simplicity. We consider *electrically charged* particles with charge $q$ in a spatially dependent electrical potential $\phi$. To solve the equations consistently, $\phi$ must be calculated from the Poisson equation,

$$\vec{\nabla} \cdot (\epsilon_0 \epsilon_r \vec{\nabla} \phi) = -\rho, \tag{4.12}$$

which contains the relative dielectric permittivity, $\epsilon_r$, and the space charge, $\rho$, which itself depends on $n$. Since space charge formation is not the focus here, we will assume, for the moment, that $\phi$ is given, and start from the continuity equations for the carrier density $n$ and energy density $u$,

$$\frac{\partial n}{\partial t} + \vec{\nabla} \cdot \vec{j}_n = 0 \tag{4.13}$$

$$\frac{\partial u}{\partial t} + \vec{\nabla} \cdot (\vec{j}_u + q\phi \vec{j}_n) = 0. \tag{4.14}$$

The term $q\phi\vec{j}_n$ takes into account that the carriers with charge $q$ carry an electrical energy $q\phi$; $\vec{j}_u$ is the remaining non-electric part of the energy current. The entropy balance equation,

$$\frac{\partial s}{\partial t} + \vec{\nabla} \cdot \vec{j}_s = \dot{s} \tag{4.15}$$

can now be derived with the help of the first law, $du = Tds + (\mu + q\phi)dn$, in the following way. According to Eq. (3.45) the *electrochemical potential* $\mu_{ec} = \mu + q\phi$ is introduced, which is the relevant (gauge-invariant) intensive quantity describing the energetic behavior of charged particles. We replace the differentials by time derivatives and substitute $\partial n/\partial t$ and $\partial u/\partial t$ from Eqs. (4.13) and (4.14) and obtain

$$\frac{\partial s}{\partial t} = \frac{\left(\frac{\partial u}{\partial t} - \mu_{ec}\frac{\partial n}{\partial t}\right)}{T} = \frac{\left(\mu_{ec}\vec{\nabla} \cdot \vec{j}_n - \vec{\nabla} \cdot (\vec{j}_u + q\phi\vec{j}_n)\right)}{T}. \tag{4.16}$$

The right side consist of terms of the type $a\vec{\nabla} \cdot \vec{b}$ with scalars $a$ and vectors $\vec{b}$, which can be written in the form $\vec{\nabla} \cdot (a\vec{b}) - \vec{b} \cdot \vec{\nabla}a$. We place the $\vec{\nabla} \cdot (a\vec{b})$-terms to the left side in order to get the form of Eq. (4.15). The remaining task is to identify the expressions for $\vec{j}_s$ and $\dot{s}$:

$$\vec{j}_s = \frac{1}{T}\vec{j}_u - \frac{\mu}{T}\vec{j}_n \tag{4.17}$$

$$\dot{s} = (\vec{j}_u + q\phi\vec{j}_n) \cdot \nabla\left(\frac{1}{T}\right) - \vec{j}_n \cdot \vec{\nabla}\left(\frac{\mu_{ec}}{T}\right). \tag{4.18}$$

Note that in the entropy current density (4.17) $\mu$ (and not $\mu_{ec}$) occurs, because $\vec{j}_u$ was defined as the nonelectric part of the energy current density. We did this since we added separately the electrical contribution. However, for being consistent with the usual notation, one should include the term $q\phi\vec{j}_n$ in $\vec{j}_U$, such that in Eqs. (4.17) and in Eq. (4.18), $\mu$ is replaced by $\mu_{ec}$ and $\vec{j}_u + q\phi\vec{j}_n$ is replaced by $\vec{j}_u$, respectively.

But let us continue here with our original notation. As usual, the entropy production rate turns out to be a sum of products of generalized current densities and forces, associated with gradients of inverse temperature and ratio of (electro-) chemical potential and temperature. We promised to re-derive Eq. (4.9). You may wonder where in Eq. (4.18) the power density $p = jE/T$ associated with the Joule heat power density $p$ is hidden (here, $j = qj_n$ is the electric current, and $E = -\nabla\phi$ is the electric field). In fact, it is hidden in the second term; differentiation and evaluation of all terms which contain $\phi$ in Eq. (4.18) gives

$$\dot{s} = \vec{j}_u \cdot \vec{\nabla}\left(\frac{1}{T}\right) - \vec{j}_n \cdot \vec{\nabla}\left(\frac{\mu}{T}\right) + \frac{p}{T}. \tag{4.19}$$

Expressions (4.18) and (4.19) represent the *local* entropy production rate density. The total rate $\dot{S}$ additionally requires a volume integration of $\dot{s}$. To show this, we consider a large isolated system, consisting of the subsystem containing the just discussed gradients and currents, embedded in an environment which is a heat bath at constant temperature $T_{amb}$, and constant chemical potential. We select the volume $V$ such that its surface $\partial V$ is deep inside the ambient reservoir with $T_{amb} = T_2$. For the stationary state $(\partial s / \partial t = 0)$, and with Eq. (4.17), one obtains

$$
\begin{aligned}
\dot{S} &= \int_V dV \, \dot{s} = \int_V dV \, \vec{\nabla} \cdot \vec{j}_s = \int_{\partial V} d\vec{A} \cdot \vec{j}_s = \frac{1}{T_{amb}} \int_{\partial V} d\vec{A} \cdot \vec{j}_u \\
&= \frac{1}{T_{amb}} \int_V dV \, \vec{\nabla} \cdot \vec{j}_u = -\frac{1}{T_{amb}} \int_V dV \, \vec{\nabla} \cdot (q \vec{j}_n \phi) \\
&= -\frac{1}{T_{amb}} \int_V dV \, \vec{j} \cdot \vec{\nabla} \phi = \frac{1}{T_{amb}} \int_V dV \, \mathrm{p} = \frac{P}{T_{amb}}
\end{aligned} \tag{4.20}
$$

where we used $\vec{j}_n = 0$ at the volume surface $\partial V$ (by definition, the heat bath can only exchange heat), and denoted the electrical current by $\vec{j} = q\vec{j}_n$. For the step in the second line, Eq. (4.14) was used.

### 4.2.2 $\dot{S}$ IN BALLISTIC TRANSPORT

From Section 3.3.2 we know that there is no entropy production in the spatial region of (stationary) ballistic transport, because the distribution function $f$ remains constant. The change of the distribution function - and thus the entropy change - occurs where the carriers (photons, phonons, electrons, ...) leave and enter the reservoirs, i.e., at the contact between reservoirs and conductor. Let us illustrate this with two examples: radiative heat transport through a vacuum (see Figure 4.2) and electric transport in mesoscopic conductors.

#### 4.2.2.1 Photons: Radiative heat transfer

We consider again the situation if Figure 3.4 b) with two blackbody heat reservoirs. Reflection of incoming radiation at the reservoir surface is neglected as usual for blackbodies. It is sufficient to look at the part of the radiation in the positive z-direction from reservoir 1 to reservoir 2. For the reverse current in the minus z-direction, an analogous consideration can be done separately. The photons with positive wavenumber $k_z > 0$ leave reservoir 1 at temperature $T_1$ with distribution $f(T_1)$. The associated heat current is given by the first term in Eq. (3.52), $j_u^+ = \sigma_{SB} T_1^4$. Remember that the polar-angle integration in Eq. (3.51) led to the factor $1/4$, such that the current density of the quantity $U/V$ did not become $cU/V$ but $cU/4V$. The same integration factor will be obtained if one calculates the *entropy* current density in the z-direction, $j_S$, which leaves reservoir 1, such that with Eq. (2.59) it holds that

$$
j_S^+ = \frac{cS}{4V} = \frac{1}{3} ca T_1^3 = \frac{4}{3} \sigma_{SB} T_1^3 = \frac{4}{3} \frac{j_u^+}{T_1}. \tag{4.21}
$$

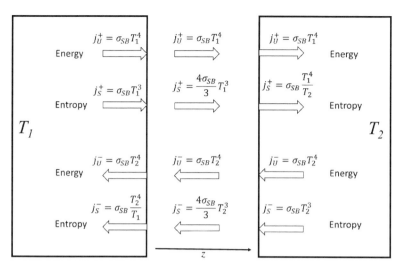

**Figure 4.2** Energy and entropy current densities for heat flow between two reservoirs at temperatures $T_1$ and $T_2$ separated by a vacuum gap. The first and third columns refer to the currents at the reservoir surface on the reservoir side, while the middle column represents the current densities in the vacuum region.

Consider now the surface of reservoir 1 as shown in Figure 4.2. There is obviously a discontinuity in the entropy current density at the surface of reservoir 1 (first subscript of $\dot{S}_{jk}$), which corresponds to a local entropy production rate associated with radiation coming from reservoir 1 (second subscript)

$$\dot{S}_{11} = \frac{4A\sigma_{SB}T_1^3}{3} - A\sigma_{SB}T_1^3 = \frac{A\sigma_{SB}T_1^3}{3}. \tag{4.22}$$

$A$ is the surface area, and $\dot{S}_{11}$ is obviously positive. On the other hand, when radiation from reservoir 1 is absorbed at the surface of reservoir 2 at temperature $T_2$, Figure 4.2 implies an entropy production rate

$$\dot{S}_{21} = -\frac{4A\sigma_{SB}T_1^3}{3} + \frac{A\sigma_{SB}T_1^4}{T_2}. \tag{4.23}$$

Analogous expressions hold for the other surface entropy production rates, $\dot{S}_{k2}$, ($k = 1, 2$), of the radiation stemming from reservoir 2. Summing up all contributions leads then to the net entropy production rate $\dot{S}_{tot} = Aj_U(T_2^{-1} - T_1^{-1})$, as one expects. The contributions from the first terms in Eqs. (4.22) and (4.23), having the factor $4/3$, cancel each other out. Each of the four terms, $Aj_U^{+/-}/T_k$, can be interpreted as entropy production rate of the current in the $+/-$-direction at the contact with reservoir $k = 1, 2$.

You may certainly wonder what it means that the $\dot{S}_{jk}$ are nonzero in equilibrium $T_1 = T_2$, while the total $\dot{S}_{tot}$ of course vanishes. In full equilibrium, you cannot measure any current or entropy production. But you are free to decompose in theory the vanishing quantities in immeasurable contributions with finite values. The separation of the vanishing equilibrium current into drift and diffusion currents is a similar thing.

A main result that should be kept in mind is that the *entropy production in ballistic transport occurs when the carriers, which carry the intensive quantities ($T$, $\mu_{ec}$, $p$, ... ) of the emitting reservoir, cross the reservoir contacts, and particularly when they enter the receiving reservoir where they equilibrate to new values of their intensive variables.*

It is a simple step now to go to more complicated systems consisting of many reservoirs coupled via ballistic transport channels. A treatment requires then the knowledge of which fractions of the carriers coming from reservoir $k$ are scattered into reservoir $l$. We will not discuss this here; but if you are interested in a nice introduction to the scattering formalism of transport (mainly for electrical systems), you will find it in the book [Dat97], which is also a main reference for the following example.

### 4.2.2.2   Electrons: Mesoscopic conductors

An example of large didactic value is quantized ballistic transport of electrons through (*mesoscopic*) conductors (low-dimensional conductors where quantum effects play a role), because it helps to the understand the nature of electric resistance. Two important realizations are electrical conduction through very narrow constrictions (e.g., quantum point contacts or certain types of carbon nanotubes) and the *quantum Hall effect*. The one-dimensional channel in the former example occurs via the effective geometric smallness of the constriction. For ballistic transport, additionally the characteristic carrier scattering length in the channel must be long as compared to the channel length. In the quantum Hall effect, the one-dimensional channels are formed due to the huge magnetic field which forces the electric carriers to move along the edge of the thin Hall sample (*edge states*) without backscattering.

Consider first the illustration of a one-dimensional electrical conductor shown in Figure 4.3 (a) [Chr06], and assume that the electron reservoirs are at the same temperature and differ only in their electrochemical potential by $\Delta\mu_{ec} = -e\mathsf{U}$, where $\mathsf{U}$ is the applied (negative) voltage and $e$ is the elementary charge. Below the lower of the two Fermi levels, there is no transport because all states are occupied in both reservoirs. It is thus sufficient to look at the current in the energy interval $\Delta\mu_{ec}$, which flows from the left to right reservoir. Equilibration of an electron happens when it arrives at the receiving reservoir, because it comes with an energy $\Delta\mu_{ec}$ above the Fermi level and will dissipate this surplus energy to the reservoir. Let us determine the power $P$, and calculate the rate at which the electrons are absorbed. Because of the Pauli principle, the electrons can only travel one by one per spin degeneracy. They come thus in a row. Let us consider a single spin state. Suppose they come as close as possible one after the other (which may be justified with a maximum

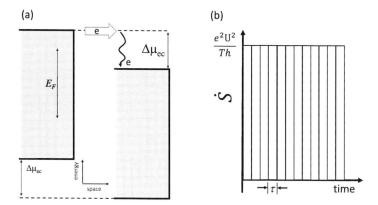

**Figure 4.3** (a) Energy diagram of two electron reservoirs with Fermi energy $E_F$ and separated by a one-dimensional ballistic channel. Application of a voltage results in a shift of the energy levels by $\Delta\mu_{el}$ and thus to a flow of electrons from left to right. They equilibrate by releasing their surplus energy $\Delta\mu_{el}$ in the right reservoir within a decay time $\tau$. The consequent entropy production rate is shown in (b) as a function of time. The fermionic character of the carriers forces them to pass through the one-dimensional channel one-by-one (for each spin state).

entropy production rate; see Section 4.3). Then the rate of transmitted electrons is limited by the time $\tau$ needed for the incoming electrons to dissipate their excess energy $\Delta\mu_{ec}$ to the reservoir. Assume that the ultimate limit of this decay time, $\tau$, is given by the energy-time uncertainty relation $\Delta\mu_{ec}\tau \approx h$. Then, $\tau \approx h/\Delta\mu_{ec}$, and the power becomes $P \approx \Delta\mu_{ec}/\tau = \Delta\mu_{ec}^2/h = e^2 U^2/h$. Because the power is related to the electrical conductance, $G_{el}$, by $P = G_{el}U^2$, one obtains the so-called *Landauer conductance quantum*

$$G_{el} = \frac{e^2}{h} \tag{4.24}$$

for the electric conductance of a one-dimensional, ballistic channel and one spin direction.

Our derivation was somewhat heuristic, and a more rigorous derivation follows along the same lines as for Eq. (3.54), but now with fermions instead of bosons. A rough explanation goes like this. The total electrical current in the channel is given by

$$I = \frac{-e(dN/dE)\Delta\mu_{ec}}{(L/v)}, \tag{4.25}$$

where the nominator contains the transferred charge, and the denominator the transfer time. Up to the factor $-e$, the charge is given by the density of states, $dN/dE$, at the Fermi energy times the energy interval $\Delta\mu_{ec}$ (cf. Figure

4.3 (a)). The transfer time is determined by the channel length $L$ and the Fermi velocity $v$. Now we need to determine the density of states $dN/dE$. In one dimension, the energy-momentum ($E$-$p$) relation $E = p^2/2m$ of a particle with mass $m$ and velocity $p$ leads to $dE = v dp$ (see also the last bullet in the list after Eq. (2.73)). The total number of states in a one-dimensional channel of length $L$ and momentum interval $dp$ becomes $dN = L\, dp/h = L\, dE/hv$, according to our arguments after Eq. (2.41), now with $dx = L$. We can substitute $dN/dE$ in Eq. (4.25) with $\Delta\mu_{ec} = -eU$ and obtain Eq. (4.24) again.

The conductance quantum is universal as it depends only on fundamental constants. If there are two equivalent spin states, a factor of 2 occurs. In general, mesoscopic conductors often consist of several one-dimensional (parallel) sub-bands, associated with the discrete quantum levels due to the transverse confinement of the electrons in the transport channel. The total conductivity is then the sum of their individual conductivities. If they have, due to some elastic scattering, a carrier transmission probability, which is lower than one, you must also multiply each of the sub-band conductivities by these factors before summing them up. We will not go more into detail, but just mention that in the quantum Hall effect the spin degeneracy is split by the strong magnetic field, and a Hall resistance $h/e^2 = 25.8\ k\Omega$ (*von Klitzing constant*) per edge state (or Landau level) is measured. The von Klitzing constant was even considered as a candidate for the definition of the unit Ohm, which is, after all, *the* unit associated with electrical losses.

## 4.3   ENTROPY PRODUCTION RATE OPTIMIZATION

Equilibrium states can be determined by solving an optimization problem, like maximizing entropy, minimizing free energy, etc.. This makes things simple. There are plenty of other problems in physics which can be solved with the help of putting derivatives or variations of functions or functionals at zero. It is thus fair to pose the question: does a variational principle also exist for the determination of nonequilibrium states of thermodynamic systems? As already said above: no, not in general! But linear nonequilibrium thermodynamic systems with a symmetric Onsager matrix are exceptions. You can find more details relevant to the following discussion in Refs. [MS06, Ich94]. Let us start with a heuristic argument for a *maximum entropy production rate principle*. You saw earlier that the extremely fast processes associated with microscopic degrees of freedom (like the molecular collisions in a gas) are responsible for the decay of nonequilibrium states to equilibrium, and thus for entropy production. One may wonder why the equilibration is (macroscopically) slow, despite of the rapidness of the microscopic equilibration mechanism acting on a very short time-scale, $t_{micr}$. Why is the entropy production rate not huge, like $\Delta S/t_{micr}$? One answer lies in the existence of constraints in the form of local conservation laws for energy, particle number, momentum, etc.. So, $\dot{S}$ cannot become arbitrarily large. It is then tempting to formulate the hypothesis that *the entropy tries to grow as fast as possible* to its maximum, where *as fast as possible* means: *maximum rate $\dot{S}$ subject to the given constraints*. Of course, it is not that simple, thus we give a more formal argument for the existence of an optimization principle under the restrictions mentioned, for weak nonequilibrium with symmetric Onsager matrix. We suppose given forces $\mathcal{F}_k$ and derive the currents $J_k$ from optimizing a function of $J_k$, while taking into account the mentioned constraints. We know that the entropy production rate is given by $\dot{S} = \sum J_k \mathcal{F}_k$, and that we need to satisfy the power balance constraint $T\dot{S} = P$, where $P$ is the irreversible power dissipation when currents $J_k$ are flowing, and $T$ is the temperature of the environment. Since $P$ must be positive for all possible non-vanishing current distributions, it must have the form

$$P = \sum_{k,l} R_{kl} J_k J_l \qquad (4.26)$$

in leading order of the $J_k$. Here $R_{kl}$ is a positive definite *dissipation matrix*. Let us now maximize $\dot{S}$ subject to the constraint by using a Lagrange multiplier $\lambda$:

$$\frac{\partial}{\partial J_k} \left( \sum_k \mathcal{F}_k J_k + \lambda \left( \sum_k \mathcal{F}_k J_k - T^{-1} \sum_{l,k} R_{lk} J_l J_k \right) \right) = 0, \qquad (4.27)$$

which gives

$$\mathcal{F}_k = \frac{2}{T(\lambda^{-1} + 1)} \sum_l R_{kl} J_l. \qquad (4.28)$$

Substitution of $\mathcal{F}_k$ in the constraint $T\dot{S} = P$ yields immediately $\lambda = 1$. Hence,

$$J_k = T \sum_l R_{kl}^{-1} \mathcal{F}_l, \qquad (4.29)$$

where $R_{kl}^{-1}$ are the coefficients of the inverse dissipation matrix. This result implies that Eq. (3.18) can be associated with an entropy production rate maximum. It holds that $L_{kl} = T R_{kl}^{-1}$. This is not a derivation of the Onsager coefficients $L_{kl}$, as we would need the $R_{kl}$-values. But we have shown the *existence* of the optimization principle. There are various slightly different ways to show this, sometimes referred to as the principle of *Onsager*, *Prigogine*, or *Ziegler* [MS06]. Depending on whether forces or currents are fixed, maximum or minimum entropy production rate principles can be applied in these special linear nonequilibrium cases. For specific applications in electrical and radiative transport, see Refs. [Chr06, CF14]. It is re-emphasized that the optimization principles discussed in this section and later in Section 8.2 correspond to different contexts and must not be confused.

# 5 Endoreversible Thermodynamics

In the previous chapters we discussed separately reversible and irreversible processes. Now we introduce *endoreversible thermodynamics* where real systems are modeled by combining reversible and irreversible subsystems. This leads then to relations between efficiency and power output as introduced in Section 1.2. Endoreversible thermodynamics goes back to work by Chambadal [Cha57] and Novikov [Nov58], and became most famous from a paper by Curzon and Ahlborn [CA75]. There exist a number of simple cases with large didactic value, because they can explain important general results for the behavior of realistic, irreversible thermodynamic engines on a rather fundamental level [HBS97, DV08]. This chapter aims to provide an easy introduction with illustrative examples.

## 5.1 ENDOREVERSIBLE SYSTEMS

Reconsider for a moment Figs. 2.10 (b) and 4.1 (a), which demonstrate, respectively, a completely reversible process (Carnot cycle) and an irreversible process (heat conduction). If you were asked to model an *irreversible* Carnot engine and forgot what we did in Section 3.1.2, you might have the alternative idea to just put the two reversible and irreversible processes of Figs. 2.10 (b) and 4.1 (a) together. This is simply what we are going to do now. In fact, an *endoreversible system* is defined as a network of fully reversible subsystems (reservoirs, Carnot-engines, etc.,) with connections where *irreversible* currents can flow. In Figure 2.10 (b), the connection between the heat reservoirs and the Carnot engine refers still to reversible heat exchange at the respective reservoir temperatures. However, you now know that real heat transport is irreversible and requires a temperature difference (or gradient). In practice, *heat exchangers* (cf. Section 8.2.2) between the working fluid of the reversible engine and the heat baths are used, where irreversibilities occur. The difference between the heat bath temperature and the upper temperature of the working fluid will turn out to play an important role for power and efficiency.

You may also adopt another point of view, which is related to the so-called *finite-time thermodynamics* [Bej96]. Reversible processes require infinitely slow processing, in order to strictly follow thermodynamic equilibrium states. This implies infinite cycle duration, $\Delta t \to \infty$, and thus zero average power output, $\dot{W} = \Delta W/\Delta t \to 0$. (Note that we use in this chapter the notation $\dot{W}$ for the work power in order to distinguish it from the power dissipation $P$ in the previous chapter, while in the literature you will often find $P$.) So, you will not get any power out of a reversible engine! You have

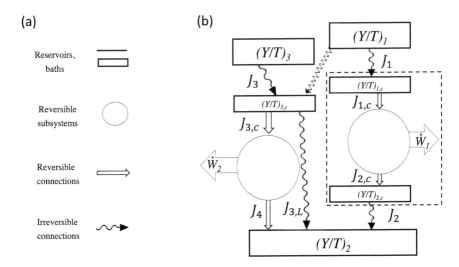

**Figure 5.1** (a) Diagrams for components in endoreversible systems. (b) Illustration of an endoreversible system with several thermal and particle reservoirs. We will mostly consider heat baths where differences in $1/T$ drive the currents, but the theory is applicable to general variables, $Y/T$.

to make the process within finite time. Imagine that in Figure 2.10 the cycle refers to a vessel of fluid which is in alternating contact with hot and cold heat baths in between the (adiabatic) working steps. In a real cycle process, these heat transfer contacts last for finite time periods (the working fluid passes through the heat exchanger in finite time). The temperature of the working fluid will then not completely equilibrate to the reservoir temperatures, and a temperature drop remains. Also the other process steps take a finite time, and what we should do is to decompose the whole cycle process in parts, using for each its own time, performing the calculations and eventually dividing the resulting work by the total cycle time. But for the moment we assume that it is sufficient to deal with constant rates (power, heat power, entropy production rate, ...), because this simplifies the calculations considerably. In the following, the rates shall be understood as averages over a process cycle: the output power is

$$\dot{W} = \frac{\Delta W}{\Delta t}, \tag{5.1}$$

where $\Delta W$ is the *net work per cycle*, and $\Delta t$ is the *total cycle time*. One could do the modeling better (in terms of time dependent ordinary differential equations for inclusion of the system's dynamic behavior), however, we will not go that far and restrict our discussions to averaged stationary fluxes.

Although many examples below will concern heat engines, endoreversible thermodynamics is not restricted to them and may include all kinds of thermodynamic forces and currents. Let us quickly outline the general theory, which is as mentioned a combination of pieces that you now know. For this, it is convenient to introduce diagrams with well-defined symbols for cycles, reversible and irreversible connections, and other components. A short possible list of diagrams is shown in Figure 5.1 (a), and an example system is depicted in Figure 5.1 (b). The reservoirs (labeled by $r = 1, 2, 3...$) are characterized by intensive variables (e.g., $T$, $p$, $\mu$, ...). The main actors are still the forces $\mathcal{F}$, which are differences in the intensive quantities $Y/T$ (e.g., $(-1/T, -p/T, \mu/T, ...)$), and drive the currents $J$ (e.g., $J_U = \dot{U}$, $J_V = \dot{V}$, $J_N = \dot{N}$, ... ) of the conjugate extensive quantities $X$ (e.g., $U$, $V$, $N$, ...). For example, the currents $J_1$ from reservoir 1 to reservoir $1, c$ (see Figure 5.1 (b)) are, within linear nonequilibrium, linear functions of the differences $(Y/T)_1 - (Y/T)_{1,c}$. The linear relation is represented by the Onsager matrix. The reversible subsystems in Figure 5.1 (b), which consist of reservoirs with subscript $c$ and a circle representing the reversible engine, can be treated with the methods of equilibrium thermodynamics. Power balance implies for the work-power output of such an engine

$$\dot{W} = (J_U)_{1,c} - (J_U)_{2,c}. \tag{5.2}$$

From the definitions of the currents $J$ (division of the $dU, dS, dX, ...$ by $dt$), one can write the first law of thermodynamics in the form

$$J_U = T J_S + \sum_{k=1}^{M} Y_k J_k \tag{5.3}$$

where we dropped the index of the reservoirs. For the *reversible* subsystems all (cycle-averaged) currents on the right-hand side of Eq. (5.3) are conserved,

$$(J_k)_{1,c} = (J_k)_{2,c} \tag{5.4}$$

including the entropy current $J_S$ (reversibility!). Last, but not least, there are relations between the currents $J$ and the forces $\mathcal{F}$ which model the irreversible connections. Within linear nonequilibrium you take the Onsager matrix, but the models can sometimes be generalized to phenomenological, nonlinear current-force relations,

$$J_k = J_k(\{\mathcal{F}_l\}). \tag{5.5}$$

For given values of the intensive parameters of the true reservoirs (those without index $c$) and considering Eqs. (5.2)–(5.5), one obtains a family of solutions with different values for the power output, $\dot{W}$, and for the overall efficiency. The family is parameterized by the intensive values of the *intermediate* reservoirs, $(Y/T)_c$. The final goal is to choose the parameter values which optimize a given objective, like maximum power output or maximum efficiency. As mentioned earlier, these two goals are often contradictory, and we will discuss other objective functions in Chapter 8.

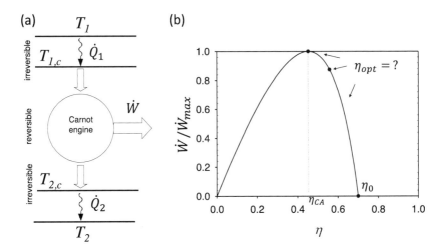

**Figure 5.2** Curzon–Ahlborn engine. (a) Endoreversible system consisting of a reversible Carnot engine with irreversible heat conduction from heat reservoirs to the working fluid. (b) Efficiency-power relation with power maximum at the Curzon–Ahlborn efficiency $\eta_{CA}$, and vanishing power at the Carnot efficiency, and at zero efficiency. What the optimum efficiency value $\eta_{opt}$ is, has to be answered by economic considerations (see Chapter 8).

## 5.2  EFFICIENCY-POWER RELATIONS

Let us discuss now some examples. The most simple is a Carnot engine with a single intermediate heat reservoir at $T_c$ (*Novikov engine* [Nov58]). Here $T_c$ just parameterizes the curve that represents the efficiency-power relation. It is a special case of the more general *Curzon–Ahlborn engine* as shown in Figure 5.2) (a), which works between the temperatures $T_{1,c}$ and $T_{2,c}$ of the working fluid which is entering from the upper and exiting to the lower heat exchangers, respectively. The temperature differences across the heat exchangers are related to the currents by

$$\dot{Q}_1 = C_1(T_1 - T_{1,c}) \tag{5.6}$$

$$\dot{Q}_2 = C_2(T_{2,c} - T_2) \tag{5.7}$$

where $C_{1,2}$ characterize the heat exchangers. Because we relate all times to the process cycle time, we have to consider these heat transfer coefficients as energy transferred per temperature difference *and per cycle time*. If you remember that the Carnot process consists not only of isothermal but also of adiabatic curves, you might ask why the irreversibility is only added to the former but not to the latter. Of course, what we do here is just a simplification, but at least in the context of finite-time thermodynamics it seems reasonable to assume that the heat transfer is the time-limiting process, since isothermal processes are slow while adiabatic processes can be very fast.

We may also consider other heat conduction laws, like Eq. (3.52) for radiation, or the *Fourier heat conduction law*, where $\dot{Q}_1 \propto (1/T_{1,c} - 1/T_1)$. Although the latter conduction law is more natural and leads to an even simpler result, we will consider (5.6) and (5.7) in accordance with the majority of the literature. In any case, the drop in temperature across the heat exchangers results in practice from the properties of the heat exchanger. It is clear that a good heat exchanger should have large $C_{1,2}$-values. But what are the optimum values of $T_{1,c}$ and $T_{2,c}$? And what if the currents of other quantities, like particles, are to be included? We will now have a look at four examples.

## 5.2.1  CURZON–AHLBORN ENGINE

We must calculate the power output $\dot{W}$. First, we denote by

$$\eta = \frac{\dot{W}}{\dot{Q}_1} \tag{5.8}$$

the true efficiency, while $\eta_0 = 1 - T_2/T_1$ is the Carnot efficiency in case of reversibility. We obtain then from

$$\dot{Q}_1 = C_1(T_1 - T_{1,c}) = \frac{\dot{W}}{\eta} \tag{5.9}$$

the temperature

$$T_{1,c} = T_1 - \frac{\dot{W}}{\eta C_1}, \tag{5.10}$$

which is smaller than $T_1$, and similarly from

$$\dot{Q}_2 = C_2(T_{2,c} - T_2) = \dot{Q}_1 - \dot{W} = \dot{W}(\eta^{-1} - 1) \tag{5.11}$$

the temperature

$$T_{2,c} = T_2 + \frac{\dot{W}}{C_2}(\eta^{-1} - 1), \tag{5.12}$$

which is larger than $T_2$. Now we can determine $\eta$ by substituting the expressions (5.10) and (5.12) in the reversible Carnot efficiency,

$$\eta = 1 - \frac{T_{2,c}}{T_{1,c}} \tag{5.13}$$

which leads to

$$(1 - \eta)\left(T_1 - \frac{\dot{W}}{\eta C_1}\right) = T_2 + \frac{\dot{W}}{\eta C_2}(1 - \eta). \tag{5.14}$$

Solving for $T_1 - T_2$ and division by $T_1$ gives with the definition of $\eta_0$

$$\eta_0 = \eta + \left(\frac{1}{C_1} + \frac{1}{C_2}\right)\frac{\dot{W}(1-\eta)}{\eta T_1}. \tag{5.15}$$

The $\dot{W}(\eta)$-relation becomes then

$$\dot{W}(\eta) = \frac{C_1 C_2}{C_1 + C_2}T_1\frac{\eta(\eta_0 - \eta)}{(1-\eta)}. \tag{5.16}$$

The derivation of the relations between power and efficiency is a main task of endoreversible thermodynamics. The function $\dot{W}(\eta)$ is illustrated in Figure 5.2 (b); it is an another realization of what was anticipated in Figure 1.1 (b) where $\eta_{max}$ was unity. It already contains half of the typical behavior of efficiency-power relations; as in Figure 1.1 (b) the behavior will become slightly different when leakage is included.

Consider first the reversible limit $\eta \to \eta_0$. Here the power vanishes, $\dot{W}(\eta_0) = 0$, as expected. For zero efficiency, $\eta \to 0$, again $\dot{W}(0) = 0$. This occurs for $T_{1,c} = T_{2,c}$, i.e., a Carnot engine without temperature drop. Since there are two zeros of $\dot{W}(\eta)$, which is positive in between, $\dot{W}$ must have a maximum. Putting $d\dot{W}/d\eta = 0$,

$$0 = (\eta_0 - 2\eta)(1-\eta) + \eta(\eta_0 - \eta) = \eta^2 - 2\eta + \eta_0 \tag{5.17}$$

with the physically relevant solution

$$\eta_{CA} = 1 - \sqrt{1 - \eta_0}. \tag{5.18}$$

In terms of the equilibrium bath temperatures, it can be written as

$$\eta_{CA} = 1 - \sqrt{\frac{T_2}{T_1}}, \tag{5.19}$$

which depends only on the temperature ratio of the upper and lower heat bath, similar to the Carnot efficiency. We will call it, as usual, the *Curzon–Ahlborn efficiency*. Putting Eq. (5.18) into Eq. (5.16), the maximum power value becomes

$$\dot{W}_{max} = C_{eff}T_1\eta_{CA}^2 = C_{eff}(\sqrt{T_1} - \sqrt{T_2})^2, \tag{5.20}$$

where $C_{eff} = C_1 C_2/(C_1 + C_2)$ is the prefactor in Eq. (5.16). An interesting and somewhat astonishing result is that the efficiency at maximum power depends only on the ratio of $T_1$ and $T_2$, but not on $C_k$. The square-root dependence on the temperature ratio is not universal; other heat conduction laws lead to other exponents or functional dependencies. Secondly, it means, that while the maximum determines the temperature values $T_{1,c}$ and $T_{2,c}$ from Eqs. (5.10) and (5.12), there is no information on the optimum $C_k$ values. Of course, they should be as large as possible. Curzon and Ahlborn considered a more differentiated model and included the different time durations $t_1$ and $t_2 = t_{tot} - t_1$ of the heat transfer processes in the upper and lower heat exchanger, and performed the calculation for the work $\Delta W$. We emphasized that the $C_k$ were defined as time averages over a cycle. If one wants to use real heat transfer coefficients, $C_k$, one must in the expression for $C_{eff}$ replace $C_k$ by $\tilde{C}_k t_k$, to obtain heats instead of heat rates. In order to eventually have the average power per cycle again, this expression must be divided by the total time $t_{tot}$ of the cycle. We can then maximize this expression,

$$C_{eff} = \frac{1}{t_{tot}} \left( \frac{1}{\tilde{C}_1 t_1} + \frac{1}{\tilde{C}_2 (t_{tot} - t_1)} \right)^{-1}, \tag{5.21}$$

as a function of $t_1$ and obtain $\sqrt{\tilde{C}_1} t_1 = \sqrt{\tilde{C}_2}(t_{tot} - t_1)$. With this one gets

$$C_{eff} = \frac{\tilde{C}_1 \tilde{C}_2}{(\sqrt{\tilde{C}_1} + \sqrt{\tilde{C}_2})^2}. \tag{5.22}$$

A remaining question concerns the entropy production rate, $\dot{S}$, of the endoreversible engine. It is simply given by

$$\dot{S} = \left( \frac{\dot{Q}_2}{T_2} - \frac{\dot{Q}_1}{T_1} \right) = \dot{Q}_1 \left( \frac{1}{T_2} - \frac{1}{T_1} \right) - \frac{\dot{W}}{T_2} \tag{5.23}$$

which can, with the help of Eqs. (5.8) and (5.16), be written as

$$\dot{S} = \frac{\dot{Q}_1}{T_2}(\eta_0 - \eta) = \frac{C_{eff} T_1}{T_2} \frac{(\eta_0 - \eta)^2}{(1 - \eta)}. \tag{5.24}$$

We will use this expression later in Section 8.3. For $\eta = 0$ (absence of a Carnot engine), the entropy production rate of normal heat transport, Eq. (4.8), is recovered. In the reversible case, $\eta = \eta_0$, there is of course no entropy production. At maximum power, $\dot{S}$ has an intermediate value.

## 5.2.2  NOVIKOV ENGINE WITH LEAKAGE

For simplicity, without restriction of the general insight, we will now continue the discussion where only an upper heat exchanger is present, i.e., $C_2 \to \infty$ and $C_{eff} = C_1 =: C$. We will now add heat leakage, $\dot{Q}_L$, to this so-called *Novikov engine* as is indicated in Figure 5.3 (a). In (b), the efficiency-power relation Eq. (5.16) is plotted as a dashed curve, where $\dot{W}$ is normalized by the maximum power $\dot{W}_{max}$ given by Eq. (5.20). If one takes leakage into account, it is most convenient to construct the graph of $\dot{W}(\eta)$ in parameterized form. This is easily done if one takes $T_{1,c}$ as the parameter, which runs from $T_2$ to

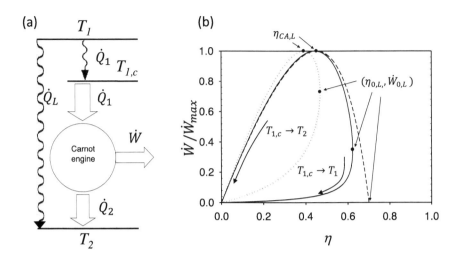

**Figure 5.3** (a) Novikov engine with additional heat leakage power $\dot{Q}_L$. (b) Normalized efficiency-power relation $\dot{W}(\eta)$ for different strength of leakage: dashed (no leakage), solid (moderate leakage), dotted (larger leakage).

$T_1$. The curve $(\eta(T_{1,c}), \dot{W}(T_{1,c}))$ is then given by

$$\dot{W}(T_{1,c}) = \left(1 - \frac{T_2}{T_{1,c}}\right) C(T_1 - T_{1,c}) \qquad (5.25)$$

$$\eta(T_{1,c}) = \frac{\dot{W}(T_{1,c})}{\dot{Q}_L + C(T_1 - T_{1,c})}, \qquad (5.26)$$

where Eq. (5.25) describes the reversible Carnot cycle between temperatures $T_{1,c}$ and $T_2$, while Eq. (5.26) is the definition of the efficiency $\eta = \dot{W}/(\dot{Q}_1 + \dot{Q}_L)$ including leakage. Figure 5.3 (b) shows the efficiency-power relations for zero (dashed curve), intermediate (solid curve), and large (dotted curve) value of $\dot{Q}_L$.

First, the power maximum value $\dot{W}_{max}$ given by $d\dot{W}/dT_{1,c} = 0$ from Eq. (5.25) is independent of leakage. The corresponding temperature of the intermediate upper bath is

$$T_{1,c} = \sqrt{T_1 T_2}. \qquad (5.27)$$

Of course, the corresponding efficiency value shifts to smaller values in case of leakage. Substitution of Eq. (5.27) and $\dot{W}_{max}$ into Eq. (5.26) yields

$$\eta_{CA,L} = \eta_{CA} \frac{\dot{W}_{max}}{\dot{W}_{max} + \eta_{CA}\dot{Q}_L}. \qquad (5.28)$$

Secondly, it is clear from Eqs. (5.25) and (5.26) that for finite $\dot{Q}_L$, both limits $T_{1,c} \to T_1$ and $T_{1,c} \to T_2$ end in the origin ($P = \eta = 0$). Consequently, the

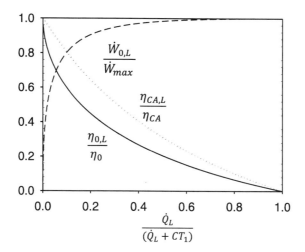

**Figure 5.4** Normalized power and efficiency values of the turning point and the power maximum, indicated by the dots in Figure 5.3, as a function of relative leakage power.

curve $\dot{W}(\eta)$ has a turning point at a maximum efficiency value $\eta_{0,L}$. The limit $T_{1,c} \to T_1$ is associated with a slowing down of the cycle process, where the efficiency decreases instead of going towards the Carnot efficiency, because the energy will get lost by leakage. The other limit, $T_{1,c} \to T_2$ is, as before, a situation where the temperature drop across the Carnot engine vanishes.

The turning point $(\eta_{0,L}, \dot{W}_{0,L})$ is calculated from the derivative of Eq. (5.26), $d\eta/dT_{1,c} = 0$, which leads to a quadratic equation in $T_{1,c}^{-1}$,

$$T_{1,c}^{-2} - 2a_1 T_{1,c}^{-1} + a_0 = 0 \tag{5.29}$$

with

$$a_1 = \frac{C}{Q_L + CT_1} \quad , \quad a_0 = \frac{CT_2 - Q_L}{Q_L + CT_1} T_1^{-1} T_2^{-1} \tag{5.30}$$

and the solution

$$T_{1,c} = \frac{1}{a_1 + \sqrt{a_1^2 - a_0}}. \tag{5.31}$$

The normalized values of $\eta_{0,L}$ and $\dot{W}_{0,L}$ as a function of $Q_L$ are plotted in Figure 5.4. The considered kind of parallel leakage reduces the efficiency, while the maximum power $\dot{W}_{max}$ remains the same ($\dot{W}_{CA,L} = \dot{W}_{CA} = \dot{W}_{max}$ should be obvious because $\dot{W}_{max}$ depends only on the bath temperatures and the heat conductance $C$). You should expect to get the same power but for less time for a given amount of primary energy.

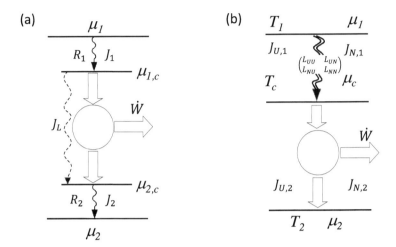

**Figure 5.5** Endoreversible (a) electrochemical engine (the leakage current $J_L$ will be discussed in Section 7.1.2); (b) thermo-electrochemical engine.

## 5.2.3 ELECTROCHEMICAL ENGINES

The formalism for endoreversible systems is general, not restricted to heat flows, and may include other currents. One could write down here the general formalism involving multidimensional currents with the Onsager matrix. But we are satisfied first with a discussion of the case of an electrochemical potential drop as shown in Figure 5.5 (a). The simplicity becomes immediately clear, if you realize that this system is equivalent to a battery in series with two resistors and an electric load power $\dot{W}$ (remember Section 1.2). The two reservoirs are then identified with the two battery electrodes, which are separated in space and thus fit to our figure. For a homogeneously mixed chemical reaction $A_1 \rightleftharpoons A_2$, the $\mu_1$ and $\mu_2$ should be associated with the chemical potentials of the educt state $A_1$ and product state $A_2$, respectively; in a homogeneous chemical system the reservoirs are not separated in space. We will come back to a battery application in Section 6.2.1. In Figure 5.5 (a), an engine which provides the power $\dot{W}$ is fed by particles which enter with electrochemical potential $\mu_{c,1}$ and exit with $\mu_{c,2}$ (although $\dot{W}$ is not produced in a Carnot engine, we stay with the index $c$). The particles originally come from and finally go to equilibrium reservoirs with electrochemical potentials $\mu_1$ and $\mu_2$, respectively. The currents $J_k$ and the potential differences $\Delta\mu_k$ are related by

$$\Delta\mu_1 \quad = \quad \mu_1 - \mu_{c,1} = R_1 J_1 \qquad (5.32)$$
$$\Delta\mu_2 \quad = \quad \mu_{c,2} - \mu_2 = R_2 J_2 \qquad (5.33)$$

where the $R_k$ are electrochemical resistances (inverse conductances). The crucial point is the conservation of the particle currents: $J = J_1 = J_2$. This leads immediately to $R_1 \Delta \mu_2 = R_2 \Delta \mu_1$, which implies

$$\mu_{c,1} = \mu_1 - R_1 J \tag{5.34}$$
$$\mu_{c,2} = \mu_2 + R_2 J. \tag{5.35}$$

The energy change of a particle equals $\mu_{c,2} - \mu_{c,1}$, such that the power as a function of the current $J$ becomes

$$\dot{W}(J) = (\mu_{c,1} - \mu_{c,2})J = \Delta \mu J - R_{tot} J^2, \tag{5.36}$$

with the total potential drop and the total resistance,

$$\Delta \mu = \mu_1 - \mu_2 \tag{5.37}$$
$$R_{tot} = R_1 + R_2, \tag{5.38}$$

respectively. We immediately obtain the maximum power as a function $J$ by solving $d\dot{W}/dJ = 0$. Since the maximum available power is $\Delta \mu J$, the efficiency can be defined by

$$\eta = \frac{\dot{W}}{\Delta \mu J} = 1 - \frac{R_{tot} J}{\Delta \mu}. \tag{5.39}$$

Solving this for $J$ and putting the result into Eq. (5.36) gives the relation we are looking for:

$$\dot{W}(\eta) = \frac{\Delta \mu^2}{R_{tot}} \eta (1 - \eta). \tag{5.40}$$

The efficiency and power values at maximum power are

$$\eta^* = \frac{1}{2} \tag{5.41}$$

$$\dot{W}_{max} = \frac{\Delta \mu^2}{4 R_{tot}}. \tag{5.42}$$

You will recognize our result discussed in Section 1.2.

## 5.2.4 THERMO-ELECTROCHEMICAL ENGINES

If there is more than a single current type, the irreversible connections can be modeled in the framework of the Onsager matrix. For energy and particle currents we did this in Section 3.2.1, see Eqs. (3.22) and (3.23). But what about the reversible engine? Look at the example in Figure 5.5 (b). The Eqs. (5.2)–(5.4) give

$$\dot{W} = J_{U,1} - J_{U,2} = \left(1 - \frac{T_2}{T_c}\right) J_{U,1} + \left(\frac{T_2}{T_c} \mu_c - \mu_2\right) J_{N,1}. \tag{5.43}$$

To arrive at this result, we used the reversibility $J_{S,c} = J_{S,2}$, particle number conservation, $J_{N,2} = J_{N,1}$, and eliminated the entropy currents. Replacement of the currents in Eq. (5.43) with the help of Eqs. (3.22) and (3.23) leads to a quadratic form of $\dot{W}$ in the forces. Optimization, by solving $\partial \dot{W}(T_c, \mu_c)/\partial T_c = 0 = \partial \dot{W}(T_c, \mu_c)/\partial \mu_c$, provides the values associated with power maximization,

$$\frac{1}{T_c} = \frac{1}{2}\left(\frac{1}{T_1} + \frac{1}{T_2}\right) \tag{5.44}$$

$$\frac{\mu_c}{T_c} = \frac{1}{2}\left(\frac{\mu_1}{T_1} + \frac{\mu_2}{T_2}\right). \tag{5.45}$$

You may ask why in Eq. (5.44) the arithmetic average of the inverse temperatures, and not the geometrical average of the temperatures as in Eq. (5.27) occurs. The reason is that we assumed here a Fourier heat conduction law (cf. Eq. (3.15)), in contrast to Eq. (5.6).

The calculation becomes more cumbersome in the presence of irreversible connections to *both* reservoirs 1 and 2. It is convenient to use then a vector notation. For our example ($J = (J_U, J_N)$, $Y = (-1/T, \mu/T)$, and **L** being the symmetric Onsager matrix with coefficients $L_{UU}, L_{UN}$, and $L_{NN}$), the bilinear form would look like this:

$$\dot{W}(T_c, \mu_c) = T_2(T_1^{-1}Y_1 - T_c^{-1}Y_c) \cdot \mathbf{L}(T_c^{-1} \cdot Y_c - T_2^{-1}Y_2). \tag{5.46}$$

Power maximization with respect to free intensive variables ($Y_c$) is straightforward. The principle should have become clear now, such that we can stop here and pass over to another interesting application case of endoreversible thermodynamics.

## 5.3 ENDOREVERSIBLE PUMPED HEAT STORAGE

The strength of endoreversible thermodynamics is to assemble large, real thermodynamic systems from simple reversible and irreversible parts. We will conclude this chapter with an example that nicely illustrates this wide-ranging applicability. Let us construct a heat storage system by connecting a heat pump with a Carnot engine, as done by Thess [The13] and shown in Figure 5.6. Exergy flow, $\dot{W}_{in}$, (work per cycle period) is first used to pump heat power $\dot{Q}_{2,L}$ from the lower heat bath 2 at $T_2$ with a heat pump (an inverse Carnot engine) to the upper heat bath 1 at $T_1 > T_2$, which receives a total heat power $\dot{Q}_{1,L}$. As for the Novikov engine, the heat transfer coefficient $C$ of the heat exchanger to the bath 1 is taken into account, hence $\dot{Q}_{1,L} = C(T_{1,L} - T_1)$; note that $T_{1,L}$ is higher than $T_1$. The stored heat is then later transformed again into power $\dot{W}_{out}$ by the Carnot engine (depicted on the right side). The heat pump and the Carnot engine are supposed to be the same, including the same heat exchanger with heat transfer conductance $C$. For clarity, the process is decomposed and illustrated in Figure 5.6 as two separate parts.

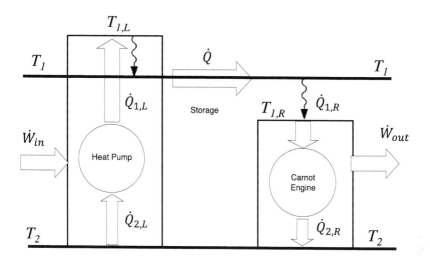

**Figure 5.6** Endoreversible thermal energy storage device, consisting of a heat pump (left, $L$), followed by a Carnot engine (right, $R$) which re-transforms the heat to work after storage.

We now calculate the overall efficiency, defined by

$$\eta = \frac{\dot{W}_{out}}{\dot{W}_{in}}, \tag{5.47}$$

of the storage device as a function of $T_{1,R}$. For this purpose, we use

$$\dot{Q}_{1,L} = \dot{Q}_{2,L} + \dot{W}_{in} = C(T_{1,L} - T_1) \tag{5.48}$$

$$\dot{Q}_{1,R} = \dot{Q}_{2,R} + \dot{W}_{out} = C(T_1 - T_{1,R}) \tag{5.49}$$

with $T_{1,L} \geq T_1$ and $T_2 \leq T_{1,R} \leq T_1$. From energy conservation $\dot{Q}_{1,L} = \dot{Q}_{1,R} \equiv \dot{Q}$ follows

$$T_{1,R} = 2T_1 - T_{1,L}. \tag{5.50}$$

Reversibility of the engine implies

$$\frac{\dot{Q}_{1,L}}{T_{1,L}} = \frac{\dot{Q}_{2,L}}{T_2} \quad , \quad \frac{\dot{Q}_{1,R}}{T_{1,R}} = \frac{\dot{Q}_{2,R}}{T_2}. \tag{5.51}$$

Solving for the heat powers yields, with Eqs. (5.48) and (5.49),

$$\dot{Q}_{1,L} = \frac{\dot{W}_{in}}{1 - T_2/T_{1,L}} \tag{5.52}$$

$$\dot{Q}_{1,R} = \frac{\dot{W}_{out}}{1 - T_2/T_{1,R}}. \tag{5.53}$$

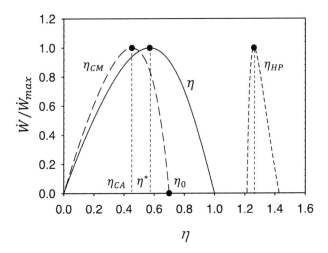

**Figure 5.7** Solid curve: normalized power output as a function of the total efficiency $\eta$ for the endoreversible thermal energy storage in Figure 5.6 with $\eta^*$ at maximum power. Long dashed: endoreversible Carnot engine ($\eta_{CM}$) with power maximum at the Curzon–Ahlborn efficiency ($\eta_{CA}$). Short dashed: efficiency (or coefficient of performance, COP) of the endoreversible heat pump ($\eta_{HP} = \eta/\eta_{CM}$).

With the help of Eqs. (5.47) and (5.50), the efficiency becomes then a function of $T_{1,R}$ only,

$$\eta(T_{1,R}) = \frac{1 - T_2/T_{1,R}}{1 - T_2/T_{1,L}} = \frac{T_{1,R} - T_2}{T_{1,R}} \cdot \frac{2T_1 - T_{1,R}}{2T_1 - T_2 - T_{1,R}}. \qquad (5.54)$$

Because $T_{1,R}$ lies between $T_1$ and $T_2$, there is only one zero of $\eta(T_{1,R})$, namely, at $T_{1,R} = T_2$. Furthermore, $\eta(T_1) = 1$, and $\eta$ is monotonously increasing between $T_2$ and $T_1$. Of course, $\eta = 1$ refers to the reversible case, where the efficiency of the heat pump is just the inverse of the Carnot efficiency. While $\eta \to 1$ for $T_{1,R} \to T_1$, the power output

$$\dot{W}_{out} = \dot{Q}_{1,R}(1 - T_2/T_{1,R}) = C(1 - T_2/T_{1,R})(T_1 - T_{1,R}) \qquad (5.55)$$

vanishes, as one expects for completely reversible processes (thermodynamic equilibrium). The power maximum, $\dot{W}_{out}(T_{1,R}) = $ max., is at the same temperature $T_{1,R} = \sqrt{T_1 T_2}$ as for the Novikov engine. The efficiency at maximum power for the thermal energy storage device is then, from Eq. (5.54), given by

$$\eta^* = \frac{2 - \sqrt{T_2/T_1}}{2 + \sqrt{T_2/T_1}}. \qquad (5.56)$$

The curve $\dot{W}_{out} = \dot{W}(\eta)$ is shown in Figure 5.7; it can be plotted by parameterizing $\dot{W}$ and $\eta$ with $T_{1,R}$. You can observe that the curve extends up to

$\eta = 1$. As mentioned, the value $\eta = 1$ is possible because the total efficiency (5.54) can be written as a product of the efficiencies of the Carnot engine and of the heat pump, $\eta = \eta_{CM}\eta_{HP}$. Since the heat pump is an inverse Carnot engine, for fully reversible operation it holds that $\eta_{CM} = 1/\eta_{HP}$. Of course, $\eta_{HP} > 1$ occurs due to its definition and the fact that the heat from the lower reservoir is a bounty; this violates neither the first nor the second law. In order to prevent confusion and to reserve the term *efficiency* for values below unity, one usually calls $\eta_{HP}$ the *coefficient of performance (COP)* of the heat pump (and, similarly, of cooling systems).

# 6 Ragone Plots

The various different energy forms require different technologies for energy storage. Needless to say that energy storage devices exist for gravitational, electric, chemical, kinetic, magnetic (inductive), etc., energy. However, despite the huge variety, a description and discussion of energy storage devices can be done on a *general* footing. To address specific details of energy storage technologies you should explore the extensive literature on this subject (e.g., [Hug10, Kul15, Ruf17]). From an application point of view, the two most relevant properties of an energy storage device are its energy capacity and the power that it can provide to a consumer or a load. The *energy-power relation* for energy storage devices, which we will call *Ragone plot* [Rag68], is a cousin of the efficiency-power relation for energy conversion devices. We will see that the two are not exactly the same. Since an energy storage device changes its state during discharge, a discussion restricted to the steady state can be inappropriate. In this chapter you should learn how to characterize an energy storage device by calculating its *Ragone plot* [CC00, Ruf17].

Ragone representations for the characterization of energy storage are used in two different contexts. The first one aims to compare different technologies in *Ragone charts*, and will only briefly be discussed in the next section; you will find much on that topic in the popular engineering literature. The second one, which will be discussed in the remainder of this chapter, is to characterize examples for *idealized* energy storage devices by their energy-power relations.

## 6.1 RAGONE CHARTS OF STORAGE TECHNOLOGIES

In order to identify an energy storage device technology for a given application, one has to know the order of magnitude of the power and of the energy that the energy storage device should be able to deliver for its task. For this purpose one identifies the associated application area in the energy-power plane. Knowing which region is covered by which energy storage technology allows to pick the one which fits best to the application. For instance, a battery for an electrical vehicle has to deliver at least an energy that is sufficient for the minimum required traveling distance. And a minimum power is required as well, that is related to the needs associated with acceleration and speed. Take a look at Figure 6.1. Batteries are in the region with high energy density, but unfortunately have comparatively low power density. This is because the energy is stored electro-chemically in the battery *volume*, where it needs some time to get the energy out, e.g., by chemistry and ion migration, which puts some limits on the power. On the other hand, capacitors provide high power, but unfortunately have limited energy. Again the reason should be obvious: the capacitor charge is stored on the electrode metal *surface* and

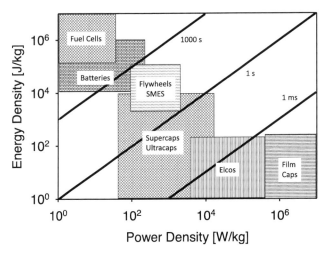

**Figure 6.1** Regions of different energy storage devices in the (specific) energy-power plane. (SMES = superconducting magnetic energy storage; *Supercaps* = super-capacitors; *Elcos* = electrolytic capacitors; *Film Caps* = polymer film capacitors).

can be quickly discharged. It is thus clear that if electrical energy is to be delivered fast, capacitors should be used, while for longer lasting energy needs, batteries are appropriate. Since the discharge time for an energy $E$ at a power $P$ is $E/P$, the lines which have a slope equal to one in the log-log plot belong to constant characteristic discharge times, as indicated in the figure.

Before the invention of super-capacitors, the gap between batteries and capacitors in the Ragone plane (in the region for discharge-times on the order of seconds) indicated that an electrical energy storage technology was missing. This pushed the development of super-capacitor technology. Super-capacitors consist of electrolyte-filled porous carbon electrodes, which combine the low thickness of the pure ionic double-layer capacitance with the huge effective electrode surface area due to the porosity. A huge specific capacitance results, and thus an energy density that is comparatively large for capacitors. The power is limited by the pore-size dependent, effective electrical resistance of the electrolyte in the pores. These circumstances exemplify the energy-power trade-off relevant for the subsequent subsections: An increased inner surface per volume (which is a design parameter) due to more and smaller pores leads to higher energy but lower power.

The application areas shown in Figure 6.1 are for illustration; in the literature similar figures with greater detail exist. To use specific energy and power values as in this figure is reasonable if the technologies scale in size. Sometimes, representations showing power or energy versus discharge time can be found, which contain the same information.

(a)　　　　　　　　　　　　　　(b)

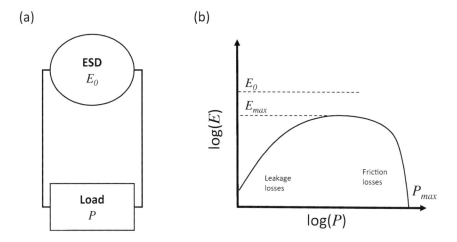

**Figure 6.2** (a) An energy storage device (ESD) with initial energy $E_0$ is discharged via a load which demands a constant power $P$. (b) The Ragone plot $E(P)$ provides the available energy $E$ for the load demanding constant a power $P$. $E$ is limited by leakage for low $P$, and $P$ is limited by internal friction and transport losses.

## 6.2　RAGONE PLOTS OF STORAGE DEVICES

Suppose an energy storage device is initially charged up with an energy $E_0$. This energy (or part of it) will be used by a load or a consumer (cf. Figure 6.2 (a)), that *demands a power* with value $P > 0$. This power is finite and can thus be delivered only during a finite time, $t_\infty$. The energy obtained by the load is $E \leq E_0$. The Ragone plot is then defined as the curve $E(P)$ in the power-energy plane. Each energy value, $E(P)$, is the available energy for *constant* power $P$.

In general you cannot expect to get out all of the initially stored energy. For very slow discharges (or very long discharge times), $P \to 0$, such that self-discharge (leakage) of the energy storage device will lead to a total loss of the stored energy, i.e., $E \to 0$. On the other hand, the power is generally limited to a maximum available power, $P_{max}$, by internal losses or other constraints (friction, internal resistance, imperfect combustion, etc.,). If the power becomes large or close to $P_{max}$, an increase of these losses leads to a drop of the available energy. There exists a maximum $E_{max}$ of $E$ in between these limits, $0 \leq P \leq P_{max}$. The situation is illustrated in Figure 6.2 (b).

The Ragone relation $E(P)$ can now be determined as follows. Let $\mathbf{q} = (q_1, q_2, ...)$ be the state variables of the energy storage device (e.g., angle coordinates, charge, current, momentum, etc.). This *state of charge* vector is in general governed by a dynamic equation

$$\frac{d\mathbf{q}}{dt} = \mathbf{F}(\mathbf{q}, t) \qquad (6.1)$$

with initial condition $\mathbf{q}(t = 0) = \mathbf{q}_0$. The initially stored energy $E_0$ is $E(t_0) = E_0(\mathbf{q}_0)$. Delivery of a finite, constant power value $P$ is only possible for finite time, $t_\infty$, which has to be calculated from the dynamic equation for the energy storage device. This will be the main challenge for determining the Ragone plot. Note that $t_\infty = t_\infty(\mathbf{q}_0, P)$ depends on the initial state, $\mathbf{q}_0$, and on the power $P$. The obtained energy, $E$, is then given by the time-integral of the power. In specific applications with time-dependent loads, $P(t)$, this means

$$E(P) = \int_0^{t_\infty} P(t)dt, \qquad (6.2)$$

which can be calculated at least numerically when Eq. (6.1) is solved. For determining the usual Ragone plot, the task is even a bit easier, because it is done for time-constant $P$. The energy-power relation becomes thus

$$E(P) = Pt_\infty(P), \qquad (6.3)$$

where we omitted the dependence on the initial condition $\mathbf{q}_0$.

The present Ragone-plot approach is not the only way to discuss energy-power relations. Alternative ways, which can have the advantage of being more easily applicable, are, e.g., the *modified Ragone representation* by Rufer and Delalay [Ruf17], and a linear response simplification for strongly frequency dependent systems [CCO99].

In the other chapters we consider the power $P$ and the efficiency $\eta$ of a device, while here it is $P$ and energy $E$. We introduce the quantity

$$e = \frac{E}{E_0}, \qquad (6.4)$$

which has a meaning similar to the efficiency. But it is not *exactly* the same thing - there are subtle differences due to two reasons. First, $e$ is only defined for the discharge process, while charging an energy storage device in general contributes also to the losses. However, the charging can in principle be optimized separately and does not depend on the $P$ of the load; therefore it is not included in the Ragone plot. For properly defining an average efficiency $\eta$ for a total cycle one must include the *charging efficiency* $\eta_{ch}$ and write $\eta = \eta_{ch}e$, provided $e$ is the discharge efficiency; the term *round-trip efficiency* is used in the literature.

Secondly, $e < 1$ does not necessarily mean that the quantity $E_0(1 - e)$ is completely lost energy. We will see below when discussing the capacitor, that some residual energy, $E_\infty$, can still be left in the energy storage device, but it cannot be used at the power value $P$; at a smaller power, however, it might be possible to harvest at least a part of it. So, $e$ is not necessarily the discharge efficiency! If you want to have a relation between Ragone plot and efficiency, you may take in the latter case

$$\eta = \eta_{ch} \frac{E}{E_0 - E_\infty}, \tag{6.5}$$

if you assume that the discharging with $P$ and the charging with $\eta_{ch}$ are only partial, i.e., between $E_0$ and $E_\infty$. Note that in general both $E$ and $E_\infty$ are functions of $P$. Equation (6.5) will become important when we investigate the capacitor.

Prior to illustrations with a few specific examples, it is helpful to have a quick look at how different classes of energy storage devices are distinguished in the framework of the Ragone-plot description. Although you may invent arbitrary types of energy storage devices (or types of dynamic systems defined by Eq. (6.1)), many can be described by an equation of the form

$$m \frac{d^2 q}{dt^2} + \frac{m}{\tau} \frac{dq}{dt} + \frac{d\Psi}{dq} = F. \tag{6.6}$$

We used parameter symbols on purpose, which you know from the mechanics of a point particle with coordinate $q$ and mass $m$, that moves in a potential $\Psi$, feels a friction force with coefficient $m/\tau$, and is subject to an external force $F$. For the general case, $m$ is the *inertia* and $\tau$ is the friction time constant. The energy can have kinetic or potential (or mixed) character, the internal losses are related to the friction, and the external load is acting with a force $F$. You certainly were taught to solve this equation, at least for simple cases. But our task is not so easy as you might think: since the load demands constant power $P$, and the power is the product of velocity and force, you have to solve Eq. (6.6) with $F = -P/\left(dq/dt\right)$. This makes the problem nonlinear. The following examples will represent special limit cases, like pure potential energy storage (e.g., batteries, capacitors, gravitational energy like pumped hydro-power, etc.,) or kinetic energy storage (flywheels, superconducting magnetic energy storage (SMES), etc.).

## 6.2.1 THE IDEAL BATTERY

Batteries store chemical energy, which they provide eventually as electrical energy to the load. Like many other storage technologies, they act thus at the same time as storage and as conversion devices. The conversion occurs in the form of electrochemical reactions at the anode where the oxidation reaction takes place, and at the cathode where the reduction reaction happens. There is a smorgasbord of different battery types, which can be classified in different ways, like primary (non-rechargeable) and secondary (rechargeable) types, batteries with aqueous (acid or alkaline) and non-aqueous electrolytes (Li-ion, NaS, ... batteries), to list a few. Again, we will not go into these de-tails but rather remain on the descriptive level of Section 5.2.3. The simplified discussion covers not only the batteries mentioned but also other electrochemical conversion devices, like flow batteries and fuel cells where the chemical substances are transported. The energy capacity of a fuel cell, where a cold and controlled oxyhydrogen reaction, $H_2 + 0.5\,O_2 \to H_2O$, takes place scales with the size of the hydrogen fuel tank.

In any case, the output of such battery-like energy storage devices is an electrical current at a certain electrochemical potential difference (the voltage). The reversible voltage of a battery, i.e., at negligible current, is given by the *Nernst equation*, which we shall now derive from our knowledge gained in Chapter 2. Since chemists like to think in units of moles, we will do so, too. For brevity, let us write the net chemical reaction between the reaction partners $A_k$ in the form $\sum \nu_k A_k \to 0$, where the stoichiometric coefficients $\nu_k$ of the educts and products are taken as positive and negative, respectively. This allows us to shuffle the right side (products) of the reaction equation to the left side (educts). Because pressure and temperature are constant, the thermodynamic potential relevant for the available work is the free enthalpy, $G$ (see Section 2.2.4). Its difference between initial and final reaction states gives us the maximum available electrical work, $\nu N_A e U_0$, where $\nu$ is the stoichiometric number of electrons obtained from the reaction, $N_A$ is the Avogadro number, and $U_0$ is the battery voltage. Let us consider, for simplicity, all ions as independent particles such that the chemical potential for the particle species $k$ has a form as discussed after Eq. (2.97). We may then write $\mu_k = \mu_k^{(0)} + k_B T \ln(a_k)$, where $\mu_k^{(0)}$ are constants associated with the values at normal conditions; they are usually different for each species. The so-called activities $a_k$ are, in the independent particle approximation, the ratios of the concentrations and standard concentration. For real (non-independent) particles, the activities are affected by the interactions - but this goes beyond our short overview. If we put these expressions with our sign convention for the $\nu_k$ in $\Delta G = -N_A \sum \nu_k \mu_k = -\nu N_A e U_0$ (see Eq. (2.90)) we obtain the Nernst equation

$$U_0 = U^{(0)} + \frac{k_B T}{\nu e} \ln \left( \prod_k a_k^{\nu_k} \right). \tag{6.7}$$

If all densities are in the standard state (i.e., all $a_k = 1$), then the cell voltage is $U_0 = U^{(0)} = e^{-1} \sum \nu_k \mu_k^{(0)}$, which can be calculated from the *standard redox potentials* for electrode reactions; the values are listed in appropriate tables. The standard potential $U^{(0)}$ is usually denoted by $E^{(0)}$, and can of course only be measured towards a second electrode which closes the electrochemical circuit; therefore one defines it against a reference electrode, the *standard hydrogen electrode*, which has by definition zero potential.

Let us consider two simple examples. For the oxyhydrogen reaction it amounts to 1.23 V; this is, of course, also the voltage required for electrolysis of pure water. For the lead-acid battery, used for instance in cars, with anodic oxidation reaction $Pb + SO_4^{2-} \to PbSO_4 + 2e^-$ with $U_a^{(0)} = $ -0.36 V and cathodic reduction reaction $PbO_2 + SO_4^{2-} + 2e^- + 4H^+ \to PbSO_4 + 2H_2O$ with

$U_c^{(0)} = 1.68$ V, one obtains a full cell standard voltage of $U^{(0)} = U_c^{(0)} - U_a^{(0)} = 2{,}04$ V.

Although we understand now the battery voltage which plays a main role for the Ragone plot, for general interest a few sentences on pure chemical reactions without electric voltage (as a special case of electrochemical reactions) are now added prior to a discussion of the Ragone plot for the ideal battery. In thermodynamic equilibrium it holds that $\Delta G = 0$, which means that the chemical potentials satisfy $\sum \nu_k \mu_k = 0$. Furthermore, the term $K = \prod a_k^{\nu_k}$ depends on pressure and temperature and is an equilibrium quantity, called the *mass action law constant*. The maximum work done by a reaction can be obtained from the free enthalpy change, $\Delta G$, and the heat from the enthalpy $\Delta H = T\Delta S = \Delta(G + TS)$ at constant $T$. Because $S = \partial G / \partial T$ at constant pressure, the reaction heat can be calculated when $\Delta G$ is known, and thus from the temperature-dependent mass action constant $K(T)$.

We consider now a given battery and describe it in a similar way as discussed already in Figure 1.1 in the introduction and in Section 5.2.3. In this chapter the symbol $Q$ will be reserved for the electric charge in the battery (SOC), and not for heat as in previous chapters. An *ideal battery* with capacity $Q_0$ (Figure 6.3 (a)) is characterized here by a constant (charge independent) reversible cell voltage $U_{bat}$ which depends on the momentaneous charge $Q \geq 0$ as follows:

$$U_{bat}(Q) = \begin{cases} U_0 & \text{if } Q_0 \geq Q > 0 \\ 0 & \text{if } Q = 0 \end{cases}. \tag{6.8}$$

In the first step, we disregard the leakage resistance $R_L$. The power is given by

$$P = UI = (U_0 - RI)I \tag{6.9}$$

where $U_0$ is the open circuit voltage, $I = -dQ/dt$ is the current, and $R$ is the internal (terminal) resistance. It must be emphasized here that $R$ does not represent only the ohmic electric resistance of the electric circuit, but contains all losses associated with the electrochemical current, and thus includes irreversible chemical or Faradaic processes. The simplification lies here in the linearity (i.e., constant $R$ and $U_{bat}$); nonlinearity will be briefly addressed below.

The solutions of the quadratic equation (6.9) are

$$I_\pm = \frac{U_0}{2R} \pm \sqrt{\frac{U_0^2}{4R^2} - \frac{P}{R}}. \tag{6.10}$$

In the limit $P \to 0$, the two branches correspond to a discharge current $I_+ \to U_0/R$ and to $I_- \to 0$. As we have seen in the first chapter, the load can also be parameterized by constant load resistance, $R_{Load}$ (this works only for the battery, but not in general, e.g., for the capacitor below). The two limits belong then to $R_{Load} \to 0$ (short circuit) and $R_{Load} \to \infty$ (open switch),

(a)                              (b)

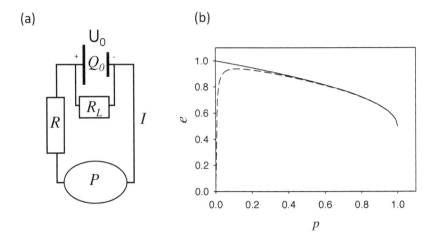

**Figure 6.3** (a) Constant power load $P$ connected to a battery with capacity $Q_0$, terminal resistance $R$, and leakage resistance $R_L$. (b) Solid curves: Ragone plot $e(p)$ of an ideal battery with (dashed) and without (solid) leakage. Here, $e$ is also the discharge efficiency, with $e(1) = 1/2$.

respectively. Hence for the Ragone plot we have to take the branch with the minus sign, $I \equiv I_-$, in (6.10).

The battery is empty ($E_\infty = 0$) at time

$$t_\infty = \frac{Q_0}{I_{tot}}, \tag{6.11}$$

where the initial charge $Q_0$ (the *capacity* of the battery) is related to the initial energy by

$$E_0 = Q_0 U_0. \tag{6.12}$$

The second step takes into account the leakage resistance $R_L$. In the denominator in Eq. (6.11) the total discharge current

$$I_{tot} = I + \frac{U_0}{R_L} \tag{6.13}$$

includes the contribution by a leakage current, which we suppose to be ohmic with leakage resistance $R_L$. Of course, $R_L \gg R$. Now, put $I_-$ from Eq. (6.10) into Eq. (6.13), and make use of Eqs. (6.11) and (6.12) to obtain from Eq. (6.3)

$$E_b(P) = Pt_\infty = \frac{2RQ_0P}{U_0 - \sqrt{U_0^2 - 4RP} + 2U_0R/R_L}. \tag{6.14}$$

This is the Ragone curve of the ideal battery. In the presence of leakage, $E_b(0) = 0$, and there exists an energy maximum at $P \approx U_0^2/\sqrt{RR_L}$. Without leakage $(R_L \to \infty)$, the maximum energy is available for vanishingly low power, $E_b(P \to 0) = E_0$. From Eq. (6.14) one concludes that there is a maximum power value

$$P_{max} = \frac{U_0^2}{4R} \tag{6.15}$$

associated with the available energy $E_0/2$ (here we neglected a small correction due to leakage). This point is the endpoint of the Ragone curve of the ideal battery, where only half of the energy is available while the other half is lost at the internal resistance. Maximum power corresponds to a fair distribution of the losses; a result which you knew already from Eqs. (1.2) and (5.42).

Let us finally express the Ragone plot for the battery in the dimensionless units $e_b = E_b/E_0$ and $p = P/P_{max}$

$$e_b(p) = \frac{1}{2} \frac{p}{1 - \sqrt{1 - p + 2R/R_L}} = \eta_b(p). \tag{6.16}$$

For the special case of a battery, $e_b$ has the meaning of a *discharge efficiency*, that equals 50% at maximum power. Ragone curves (6.16) with and without leakage are shown in Figure 6.3 for illustration. By the way, you can easily show that $e_b$ without leakage $(R_L \to \infty)$ is the same as $\eta(p)$ from Eq. (1.2), which can be written in the form $p = 4\eta(1 - \eta)$. Just plug this $p$ into Eq. (6.16) and see.

Real batteries have of course a more complicated $U_{bat}(Q, I)$-dependence than ideal ones, Eq. (6.8). The voltage can be reduced by polarization losses due to an activation overpotential associated with electrode charge transfer, and a concentration overpotential associated with a concentration drop near the electrode of the diffusion limited chemical species. Similarly, also $Q$- and $I$-dependent losses, e.g., in the form of $R = R(Q, I)$-relations, must be taken into account. For example, in the above-mentioned lead acid battery reaction, where the ions $H^+$ and $SO_4^{2-}$ of the electrolyte are consumed during discharge, the electrolyte resistance $R$ increases when $Q$ decreases. In general, the ordinary differential equation $dQ/dt = -I$ must then be solved with $Q$-dependent $I$. If an additional time dependence, e.g., via chemical and/or thermal effects, comes into play, the problem becomes more complicated, but remains conceptually analogous. For instance, if self-heating is relevant, a heat equation for the temperature can be added and solved together with the electrical and chemical equations. How to determine Ragone plots in case of time dependence should also become more clear with the following examples.

## 6.2.2  KINETIC ENERGY STORAGE

The battery from the previous example stores chemical energy and is thus of the potential energy type, similar to the capacitor, which will be discussed in the next section. This section considers a *kinetic energy storage device*. For instance, the energies stored in flywheels and superconductive magnetic energy storage devices are of the form $W = \mathcal{J}\omega^2/2$ and $W = LI^2/2$, respectively. Here, $\mathcal{J}$ is the moment of inertia, $\omega$ the angular velocity, $L$, the inductance, and $I$ the superconductive current. In the following we consider - as a prototype system that is formally equivalent to other kinetic energy storage devices - the massive point particle mentioned in connection with Eq. (6.6), with mass $m$, velocity $v$, and initial kinetic energy

$$W(t = 0) = E_0. \tag{6.17}$$

One may easily add more complicated terms to discuss more real kinetic energy storage devices, but we focus here as usual on the essential behavior.

Most important is the leakage mechanism for kinetic energy, which is given by the friction. A flywheel, for instance, will make a large number of rotations during energy release, and even if friction is negligibly small during idling, it may become significant when the energy storage device is coupled to the load. Friction forces are generally a function of the velocity $v$. For simplicity, we consider a Stokes friction force $F_f = -mv/\tau$ as indicated in Eq. (6.6), with large relaxation time constant $\tau$. A simple diagram of a discharging kinetic energy storage system is shown in Figure 6.4 (a). The dynamic equation for the velocity degree of freedom is given by Newton's equation which includes the force $-F_P$ (negative sign!) through which the load acts on the mass for the energy harvesting at constant power $P$. Since $P = vF_P$, it holds that

$$m\frac{dv}{dt} + m\frac{v}{\tau} = -\frac{P}{v}. \tag{6.18}$$

Although this equation is nonlinear, it can easily be transformed into a linear ordinary differential equation for the new variable $W = mv^2/2$, which is of course the kinetic energy. Indeed, after multiplication with $v$, Eq. (6.18) can be written as

$$\frac{dW}{dt} + 2\frac{W}{\tau} = -P. \tag{6.19}$$

It has the solution

$$W(t) = \left(E_0 + \frac{\tau P}{2}\right) e^{-2t/\tau} - \frac{\tau P}{2} \tag{6.20}$$

with $W(t = 0) = E_0$. Obviously, $W(t) < 0$ for large times. The energy storage device is thus empty at a finite time. Solving $W(t) = 0$ for $t$ yields

$$t_\infty(P) = \frac{\tau}{2} \ln\left(1 + \frac{2E_0}{\tau P}\right). \tag{6.21}$$

This is the result that we need for the Ragone plot, $E_{kin} = Pt_\infty$, the available kinetic energy. We express the result in dimensionless form, i.e., in terms of

(a)                                    (b)

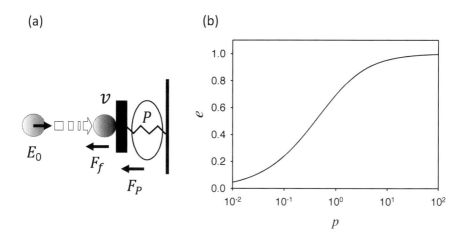

**Figure 6.4** Kinetic energy storage device. (a) A mass with initial kinetic energy $E_0$ is decelerated with a force $F_P$ by the load, which extracts the power $P$, and with a small additional friction force $F_f$. (b) Normalized Ragone plot (a power limiting mechanism exists in real cases, but is not included in our simple model).

$e_{kin} = E_{kin}/E_0$ and $p = P/P_0$,

$$e_{kin}(p) = p \ln\left(1 + \frac{1}{p}\right),$$  (6.22)

where $P_0 = 2E_0/\tau$. This curve is shown in Figure 6.4 (b). Obviously, $e_{kin} \to 0$ for $p \to 0$, which is associated with leakage due to friction at slow discharge rates. Furthermore, $e_{kin} \to 1$ for $p \to \infty$, because the friction is then irrelevant. This limit is, of course, not very realistic. There is always a mechanism which restricts the power in a real discharge and leads to losses. In the worst case it is a destructive process, like the reduced energy transfer in the collision of two equal masses due to the inelastic contribution which increases with kinetic energy. An example for a Ragone plot with a power limit is the superconducting magnetic energy storage device with ohmic bypass discussed in Ref. [CC00].

### 6.2.3   THE IDEAL CAPACITOR

We discuss now the Ragone plot for a leakage-free capacitor with capacitance $C$ and terminal resistance $R$ shown in Figure 6.5 (a). This example is of a particular didactical value, because you might think you know how a capacitor discharges, but you might not know how to derive its Ragone plot. Furthermore, the capacitor is a nice illustration for a case where $t_\infty$ is not related to the *empty* energy storage device, but the discharge may end with some

(a)                  (b)

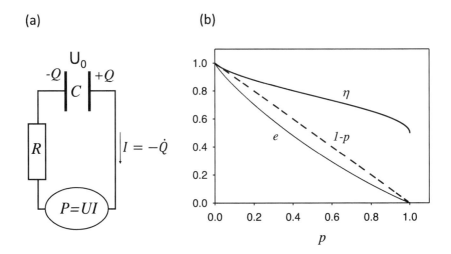

**Figure 6.5** (a) Constant power load $P$ connected to a capacitor with capacitance $C$ and terminal resistance $R$. (b) Ragone and efficiency-power plots of the capacitor without leakage (dashed curve: $e = 1 - p$).

residual charge, because the energy is not available at the *required power* $P$. From Figure 6.5 one has for the voltage at the load

$$\mathsf{U} = \frac{P}{I} = \frac{Q}{C} - RI \tag{6.23}$$

where the current $I = -dQ/dt$ is defined as the negative rate of change of the capacitor charge $Q$. According to their definitions, $Q$, $I$, and $\mathsf{U}$ are all positive during the discharge. Time differentiation of both expressions for $\mathsf{U}$ in Eq. (6.23), using $dI/dt = -(d\mathsf{U}/dt)P/\mathsf{U}^2$ for the elimination of the current, and multiplication with $\mathsf{U}$, leads to the following ordinary differential equation for $\mathsf{U}^2$:

$$\left(1 - \frac{RP}{\mathsf{U}^2}\right)\frac{d\mathsf{U}^2}{dt} = -\frac{2P}{C}. \tag{6.24}$$

This equation can be solved by separation of the variables $t$ and $\mathsf{U}^2$ and subsequent integration, which gives

$$t(\mathsf{U}) = \frac{C}{2P}\left(RP\ln(\mathsf{U}^2/\mathsf{U}_0^2) + \mathsf{U}_0^2 - \mathsf{U}^2\right). \tag{6.25}$$

Equation (6.25) should provide $\mathsf{U}(t)$, but it gives $t(\mathsf{U})$, which turns out to have a maximum. In other words, there is a turning point of the solution, given by $dt/d\mathsf{U} = 0$. You see this already from Eq. (6.24), because it happens when the term in front of $d\mathsf{U}^2/dt$ vanishes. This is the instant when the capacitor is no longer able to provide the required power $P$, and thus defines $t_\infty$. One finds for the corresponding voltage

$$U_\infty = \sqrt{RP}. \tag{6.26}$$

From Eq. (6.25) one obtains then $t_\infty$. The capacitor is not empty at this time but stops at a finite current $I_\infty = \sqrt{P/R}$, contains a residual charge $2C\sqrt{RP}$, and thus an energy

$$E_\infty = 2RCP. \tag{6.27}$$

All these relations follow directly from Eq. (6.23). In order to calculate the Ragone plot, one has to take into account that $U_0$ in Eq. (6.25) is the voltage $U$ at $t = 0^+$ when the current flows, which is not equal to the voltage $U_{C,0}$ associated with capacitor energy at $t = 0^-$ when no current is flowing yet, because there is an additional voltage drop at the resistor $R$ for $t > 0$. The capacitor voltage at $C$ equals

$$U_C = U + RI = U + \frac{RP}{U} \tag{6.28}$$

which gives

$$U_0 = \frac{U_{C,0}}{2} + \sqrt{\frac{U_{C,0}^2}{4} - RP}. \tag{6.29}$$

This must be inserted in the Ragone expression,

$$E_{cap}(P) = Pt_\infty = \frac{C}{2}\left(RP\ln(RP/U_0^2) + U_0^2 - RP\right). \tag{6.30}$$

Equation (6.29) implies the maximum power value

$$P_{max} = \frac{U_{C,0}^2}{4R}, \tag{6.31}$$

an expression which should now be familiar to you.

While for $P \to 0$, the obtained energy is $E_{cap}(0) = E_0 = CU_{C,0}^2/2$, the energy vanishes for $P \to P_{max}$. This can be understood from the voltage-dependent maximum power and the fact that the voltage drops as a function of time. The Ragone curve of the capacitor, in dimensionless units $e_{cap} = E_{cap}/E_0$ and $p = P/P_{max}$,

$$e_{cap}(p) = \frac{1}{4}\left((1 + \sqrt{1-p})^2 - p - p\ln((1 + \sqrt{1-p})^2/p)\right) \tag{6.32}$$

is plotted in Figure 6.5 (b).

Since there is an energy $E_\infty = 2RCP$ remaining on the capacitor, $e_{cap}(p)$ cannot be identified with the efficiency-power relation. We need rather to make use of Eq. (6.5). In our normalized units $e_\infty = 2RCP/E_0 = p$, such that

$$\eta_{cap}(p) = \frac{e_{cap}(p)}{1-p}. \tag{6.33}$$

You can easily show that $\eta_{cap}(1) = 1/2$, as for the battery. This had to be expected, because near $P = P_{max}$ the discharge is done at roughly constant capacitor voltage, such that the behavior is battery like. The efficiency-power relation is shown also in Figure 6.5 (b).

The three examples discussed in this chapter are very simple realizations of Eq. (6.6). In general the Ragone curve for this equation can be solved only numerically. Nevertheless, our special examples should be sufficient for understanding the concept, and you should now be able to treat more complex cases.

# 7 Power and Efficiency Limits

This chapter discusses some typical examples of performance limits in power and/or efficiency of energy conversion systems. The application examples cover solar thermal and photovoltaic energy harvesting, transformers, converters, electro-mechanical devices like motors or generators, and fluid-flow power like wind and hydro turbines. The limits are often naturally characterized by efficiency-power relations. Two sections also touch on maximum power point tracking and impedance matching. Again, the examples are reduced to their essence and are kept very simple. If you are interested in more specific applications, the literature cited below will provide plenty of additional information.

## 7.1 SOLAR POWER

Heat radiation from the sun is the main primary energy source for life on earth (another source is geothermal heat, mainly due to the decay of natural radioactive nuclei, like Uranium 235, and the cooling of the hot earth's core). There is an average heat flux of about 1.4 kW/m$^2$ from the sun arriving at the outer atmosphere of our planet; the more exact value of 1.367 kW/m$^2$ refers to the *solar constant*. Absorption and reflection from the atmosphere reduce the received power flux on the ground by about $20-30\%$. There are two main technological pathways to harvest this energy directly, namely, with *solar thermal* and with *photovoltaic* energy conversion. It turns out that the endoreversible thermodynamics of Chapter 5 provides a convenient framework for a treatment of both of them. We will see, that the two endoreversible engines for solar energy harvesting are a thermal engine for solar thermal, and an electrochemical engine for photovoltaic power. The upper heat and particle reservoir is the sun, which radiates photons from its surface at a temperature of approximately $T_1 \approx 5800\,K$. The lower bath, in a terrestrial application, is assumed to be at the ambient temperature, $T_2 \approx 300\,K$. The Carnot efficiency of an associated heat engine is then $\eta_0 = 1 - T_2/T_1 = 0.95$. Real efficiencies of solar thermal and of photovoltaic devices are, however, significantly smaller. The following subsections review a number of the concepts for discussing such solar energy efficiencies. For more details see, e.g., Refs. [Bej88], [DV00], [DV08], and [Mac15].

### 7.1.1 SOLAR THERMAL POWER

In solar thermal power conversion, heat radiation, $\dot{Q}_1$, from the sun is collected by an absorber with an effective surface area $A$. This heat power is used to produce work power, $\dot{W}$, by a heat engine. It is thus obvious to model

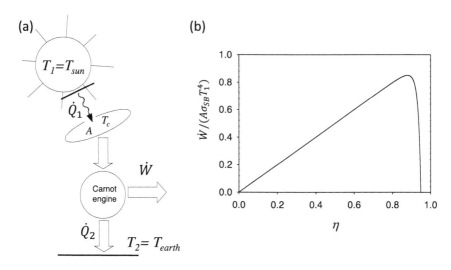

**Figure 7.1** (a) Endoreversible Novikov-type model of a solar thermal power engine, where the heat current from the upper reservoir has a Stefan–Boltzmann form. (b) The associated normalized efficiency-power relation.

this system with a Novikov engine in the framework of endoreversible thermodynamics, where the irreversible heat conductor describes heat radiation through the vacuum of outer space. Figure 7.1 (a) depicts a simple illustration. The absorber temperature plays the role of the intermediate bath temperature, $T_c$. We assume that the expression Eq. (3.52) for the heat current density between two blackbodies at temperatures $T_1$ and $T_c$ can be used. The heat current is then given by the Stefan–Boltzmann law,

$$\dot{Q}_1 = A\sigma_{SB}(T_1^4 - T_c^4), \tag{7.1}$$

where $\sigma_{SB}$ is the Stefan–Boltzmann constant. One could write a longer story on Eq. (7.1) by taking into account the absorber technology, geometry, enclosures, reduction of the radiation in the atmosphere and concentration with mirrors and/or lenses (if we talk about terrestrial systems). We will go a little bit into the geometrical issue when we discuss photovoltaic conversion below, because there - in contrast to solar thermal technology - radiation is usually not concentrated and the given geometry is a limiting factor for the incoming flux. However, here we will hide all this stuff in an *effective* surface area $A$ of the absorber (including even the fact that due to the atmospheric absorption the radiation does not exhibit an exact Planck spectrum). But all this is less relevant for the following basic discussion. The main point is that we now have a $T^4$-temperature dependence of the radiation law, which changes the form of the efficiency-power relation.

In Chapter 5 you learned how to determine the absorber temperature $T_c$

at maximum power output $\dot{W}$: the power is first expressed as the product $\dot{W} = \eta \dot{Q}_1$ with the true efficiency $\eta = 1 - T_2/T_c$. Then, by elimination of $T_c$ and $T_2$ in $\dot{Q}_1$ of Eq. (7.1) with the help of $\eta$ and $\eta_0$, the efficiency-power relation can be written as

$$\dot{W}(\eta) = A\sigma_{SB}T_1^4\eta\left(1 - \frac{(1-\eta_0)^4}{(1-\eta)^4}\right). \qquad (7.2)$$

The result is plotted in Figure 7.1 (b). A consequence of the $T^4$-dependence is that the maximum of $\dot{W}(\eta)$ must be calculated numerically. One finds $T_c \approx 2500\ K$, associated with an efficiency of $\eta \approx 0.88$. Of course, this result does not make much sense in practice, because you would hardly find an absorber material that withstands such a high temperature.

In the literature one can find a number of other ways to define optimum efficiencies associated with solar energy. The extensive discussion in Ref. [Bej88] shows which efficiency is to be considered in a given context. We will confine ourselves to a simple example. The efficiency is often defined by replacing the denominator $\dot{Q}_1$ in the ratio $\dot{W}/\dot{Q}_1$ by the *incoming* heat flux $A\sigma_{SB}T_1^4$ instead of the net flux $\dot{Q}_1$ between the upper reservoir and absorber. After division of Eq. (7.2) by this term, the resulting efficiency, $\eta_{er}$, can be written as a function of $T_c$,

$$\eta_{er}(T_c) = (1 - \frac{T_2}{T_c})\left(1 - \frac{T_c^4}{T_1^4}\right), \qquad (7.3)$$

which is the product of the Carnot efficiency $\eta_0$ of the reversible engine and the normalized net heat influx. This function has a maximum (because it is proportional to Eq. (7.2)), which is again shown in Figure 7.2. In the same figure, the true Carnot efficiency, $\eta_c = 1 - T_2/T_c$, is also plotted. The efficiencies $\eta_c$ and $\eta_{er}$ coincide at low $T_c$, but differ strongly at high $T_c$.

In the Carnot limit, $T_c \to T_1$, the heat engine runs between $T_1$ and $T_2$, without any irreversible losses by heat conduction. You understand now that the efficiency, $\eta_{er}$, which refers to the normalized power output of the endoreversible engine, vanishes in this reversible case, as well as it does in the limit $T_c \to T_2$ where all the heat is dissipated by irreversible heat conduction from the hot to the intermediate reservoir without a temperature drop at all across the Carnot engine. It is reasonable that the endoreversible efficiency is always below the Carnot efficiency, $\eta_{er} \leq \eta_0$, which needs no further explanation. Sometimes, in the literature another efficiency is discussed in the context of radiation, namely, the *Landsberg* or *Petela−Landsberg efficiency*. For the engine in Figure 7.1 (a) it is often derived by calculating the power that one obtains from the Carnot engine when one assumes the pseudo-reversibility condition

$$J_{S,1} = \frac{4}{3}A\sigma_{SB}(T_1^3 - T_c^3) = \frac{\dot{Q}_2}{T_2}. \qquad (7.4)$$

This relation prescribes a balance of the entropy current of the radiation and the entropy current into the environment. For a review of radiative entropy

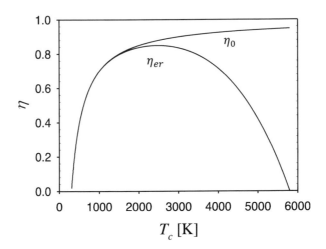

**Figure 7.2** The efficiencies of the endoreversible solar thermal engine, $\eta_0$ (Carnot) and $\eta_{er}$ given by Eq. (7.3).

currents, you may go back to Figure 4.2. We will not use Eq. (7.4) further on, because it would deserve a critical discussion in the context of the entropy production terms (4.22) and (4.23) at the surface of the (absorber) reservoir. We will rather derive the Landsberg efficiency from the exergy of radiation, and ask for the maximum available work $\Delta W$ from the change of state of a photon gas, which is initially at temperature $T_1$ (or $U_1$) in a volume $V_1$, and finally at $T_2$ (or $U_2$) and $V_2$. We learned how to calculate the exergy, $\Delta W$, by applying Eq. (2.121). With the convention of defining the efficiency with respect to the incoming flux (instead of the net flux), one defines then the maximum efficiency for this process by $\Delta W_0/U_1$, where $U_1$ is the inner energy of the photon gas in state 1. Let us rewrite Eq. (2.121) in the form

$$\Delta W_0 = U_1 - U_2 + p_2(V_1 - V_2) - T_2(S_1 - S_2), \tag{7.5}$$

where we assume that the final temperature $T_2$ and pressure $p_2$ are equal to the ambient values $T_0$ and $p_0$. The expressions for $U$, $S = 4U/3T$, and $p = U/3V$ are obtained from Eqs. (2.54), (2.59), and (2.106), respectively. Putting them into Eq. (7.5), one obtains (note that $U_2 + p_2V_2 - T_2S_2 = 0$)

$$\eta_L = \frac{\Delta W_0}{U_1} = 1 + \frac{p_2V_1}{U_1} - \frac{T_2S_1}{U_1}, \tag{7.6}$$

which can be written as

$$\eta_L = 1 - \frac{4T_2}{3T_1} + \frac{1}{3}\left(\frac{T_2}{T_1}\right)^4. \tag{7.7}$$

If you compare $\eta_L$ with the Carnot efficiency $1 - T_2/T_1$, you see that $\eta_L < \eta_0$, which is of course compulsory. The various definitions of different solar power efficiencies led to some controversy in the literature, which was discussed and resolved, e.g., by Bejan [Bej88].

## 7.1.2  PHOTOVOLTAIC POWER

The *theoretical* maximum efficiency for pure photovoltaics is considerably smaller than for solar thermal power conversion. *Pure* means here that it is exclusively the *chemical potential* carried by the photons that contributes to the harvested energy. Let us quickly describe in a simple way how a photo-diode works. If you want to learn more, you may find a nice introduction into the physics of solar cells in Ref. [WW09]. You certainly know that electrical conduction in semiconductors is based on the presence of quasi-free electrons (e) or holes (h) in the conduction band or the valence band, respectively. The creation of an e-h *pair* requires at least an energy $E_g$, associated with the energy gap between the top of the valence band and the bottom of the conduction band. The necessary energy can be provided either by phonons (thermally) or photons (optically), which kick-out electrons from filled bands, and leave back holes. Normally, the temperature is too low to have a sig-nificant number of thermally created e-h pairs. In photovoltaic devices the optically generated e-h pairs play the main role. Furthermore, carriers can also be present due to dopants, i.e., impurity atoms with a surplus electron (donors) or hole (acceptors) having energy levels so close to the band edges (some tens of meV), that they *are* thermally liberated. Dopants play an im-portant role for the creation of the internal electric field which separates the optically generated e-h pair. For this one may use a diode structure as illus-trated in Figure 7.3. By the diffusion of the electrons from the n-doped region and the holes from the p-doped region across the contact to the other sides, an equilibrium electric field forms which is governed by the balance of diffusion and drift currents. Note that the photo-voltage is not related to this diffusion potential, which is necessary for separating and guiding the *photo-generated nonequilibrium electrons and holes* to the different sides of the junction. The photo-voltage, or the photo-current, is a *bandgap effect*. An obvious condition for large efficiency is that many of the incoming photons are absorbed in the active interface region but not in the semiconductor bulk. This is one reason why in Figure 7.3 the n-region behind the transparent front electrode is thin while the p-region may be thicker.

Let us now determine the efficiency of such a photo-diode. We will ap-proximate the radiation spectrum by a Planck distribution for illustration. In fact, the true distribution is different because of the absorption and scatter-ing of the light in the atmosphere. In quantitative more exact evaluations of terrestrial photovoltaic panels, this simplification must be corrected by using the appropriate standard spectrum on earth. These so-called *Air Mass* (AMx) spectra are characterized by a number $x$ which indicates the path length of

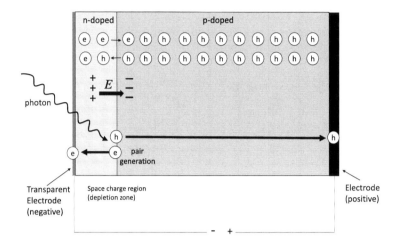

**Figure 7.3** Illustration of a photodiode. The internal electric field $E$ in the depletion region of a pn-junction is used to separate the electron and hole of a pair generated by a photon in the active region.

the light rays through the atmosphere. Outside the atmosphere one has $x = 0$, and $x = 1$ is defined on the surface of the earth for $\delta = 90°$ degrees incident angle. The effective solar constant is already considerably reduced from $x = 0$ to $x = 1$; the exact value depends on air pollution, humidity, etc.. If the angle $\delta$ is smaller, the geometrical path is longer and $x \approx 1/\cos(\delta) > 1$. Usually, efficiencies are determined for the AM1.5 spectrum, which belongs to $\delta \approx 48.2°$, and an effective solar constant of roughly $1 \text{ kW/m}^2$.

There are several contributions to the total solar cell efficiency. Each is due to a specific loss mechanism like internal electric resistance, different e-h recombination channels, reflection from the surface, and so on. Similar to heat engines, one can distinguish between two different loss types. The first corresponds to fundamental limits, while the second is associated with losses that are usually present but could at least in principle be made arbitrarily small. For instance, if ohmic losses are an issue, they might be reduced by using materials with higher electric mobility. We will focus here on two fundamental contributions to the losses. The first one is related to the fact that photons with energy below $E_g$ cannot create e-h pairs at all and are thus lost. The second one is related to *radiative recombination* of a fraction of the generated e-h pairs. This loss of carriers is unavoidable because of *detailed balance* (see Section 3.1.3): since we consider the process of e-h generation by absorption of a photon, we must also take into account the reverse process of photo-emission by e-h recombination. A detailed discussion of this and other aspects was conducted by W. Shockley and H. Queisser [SQ61]; the associated efficiency limit is called the *Shockley–Queisser limit*. Let us now partly follow their path.

Suppose that photons with energy $E = \hbar\omega < E_g$, do not interact with the semiconductor, pass directly to the environment, and are lost. But photons with $E = \hbar\omega \geq E_g$ are assumed to have a quantum efficiency of one, i.e., they create one e-h pair with energy $E_g$. This is also an approximation, because the surplus energy could be carried away by the photon or the liberated carriers, which may create further e-h pairs. There are solar cell technologies, like tandem cells or multi-gap cells, which enhance the *quantum efficiency* of a photon (i.e., number of electrons created by a photon), by making use of these energies; we refer to the designated chapter in Ref. [DV00] and will not discuss it here. The assumption of a complete loss of the surplus kinetic energy of e-h pairs is not as far-fetched as it might seem, because the carriers usually quickly thermalize via excitation of lattice vibrations (phonon emission) to the optima of the energy-bands. For fixed bandgap $E_g$, we determine the efficiency

$$\eta = \frac{\dot{W}}{J_U^+}. \tag{7.8}$$

by using Eqs. (5.43) and (3.51). We write for the denominator of Eq. (7.8)

$$J_U^+ = K_{in} \int_0^\infty f(E, T_1, 0) E^3 dE \tag{7.9}$$

with a constant $K_{in}$, and $f(E, T, \mu)$ being the Bose–Einstein distribution which depends on energy $E$, temperature $T = (T_1)$, and chemical potential $\mu(= 0)$. The term $E^3 dE$ comes from the $kd^3k$-term in Eq. (3.51), and the dispersion relation $E = \hbar\omega = \hbar ck$ of the photons. The chemical potentials of the *equilibrium* photons in the outer equilibrium reservoirs (sun and environment) are zero, hence $\mu_1 = \mu_3 = 0$. But this is different for the photon gas in the absorber. The absorber refers to the active-region at the pn-junction, which can be seen to contain a gas of electrons and holes with a temperature $T_c$ and chemical potential $\mu_c$. To understand why there is a nonequilibrium photon distribution ($\mu_c \neq 0$), you just have to conceive what happens with the Planck distribution function $f(E)$ in the junction: it remains unchanged for $E < E_g$ (of course, restricted to those $\vec{k}$ directions of incoming photons), but changes for $E > E_g$ due to the interaction of the photons with the matter. If one assumes equilibrium of the photon gas for $E > E_g$ with the e-h gas, these two thermodynamic subsystems will have the same electrochemical potentials (because photons have no charge, chemical and electrochemical potentials are the same for them). You should be aware of the simplifications. The description of nonequilibrium states by thermodynamic quantities ($T_c$ and $\mu_c$) is one simplification. The reduction to a purely chemical current is another one; it is equivalent to saying that the temperatures of the e-h gas in the photo-cell and of the environment are the same, $T_c = T_2$. The advantage of this approximation is that Eq. (5.43) simplifies to

$$\dot{W} = \mu_c J_{N,1}. \tag{7.10}$$

Now, we must be a bit careful in determining the photon current $J_{N,1}$ because of two peculiarities of photons. First, as mentioned, photons with $E < E_g$ are going directly to the environment. Secondly, the number of photons is not conserved in case of interaction with matter. One may use the argument, that the number of photons can be considered constant, because those with $E < E_g$ do not interact with the matter, and those with $E > E_g$ just loose an energy $E_g$ and continue their travel to the environment, without being eliminated (the number of photons with $E = E_g$ is of measure zero). But now comes detailed balance: some of the e-h pairs will recombine by emitting a photon (there are also other possible recombination mechanisms, but we disregard them). Because they are emitted by the e-h gas, they have temperature $T_c = T_2$ and chemical potential $\mu_c$ (cf. Figure 5.5 (b)). The net photon current which generates active e-h pairs becomes then

$$J_{N,1} = K_{in} \int_{E_g}^{\infty} f(E, T_1, 0) E^2 \, dE - K_{out} \int_{E_g}^{\infty} f(E, T_2, \mu_c) E^2 \, dE, \tag{7.11}$$

where we subtract the part associated with recombination. The values of $K_{in}$ and $K_{out}$ differ: the former depends on how the radiation is focused or concentrated towards the photo-cell, while $K_{out}$ depends more on the cell geometry itself. We will come back to the ratio $K_{in}/K_{out}$ later.

What are the meaning and value of the electrochemical potential, $\mu_c$? $I = eJ_{N,1}$ is the electric current, since $J_{N,1}$ is the net number of photons per time which produce each one e-h pair. (You might want to have a factor of 2, something like $I = 2eJ_{N,1}$, because *two* carriers are generated by one photon. But the electron and hole move in series rather than in parallel, each contributing to a single side of the p-n-junction; so there is no factor 2.) Furthermore for the electrical power, one has $\dot{W} = I\mathsf{U}$ with electric voltage $\mathsf{U}$ of the cell, such that Eq. (7.10) implies $\mu_c = e\mathsf{U}$. For completeness, one should also mention that the absolute value of $K_{in}$ is not only proportional to the incoming photon flux, but also to the volume $lA$ of active e-h pairs, where $l$ is not just the thickness of the field region of the p-n-junction, but contains also the e- and h-diffusion lengths on either side (which are typically larger than the field-region). The $K_{out}$-contribution to $J_{1,N}$ can be interpreted as a leakage loss $J_L$.

We can now determine from Eqs. (7.8)–(7.11) the maximum efficiency value $\eta$ as a function of $\mu_c$ at a *fixed* value of the bandgap $E_g$. Because it is the same thing to maximize the power $\dot{W}$ as a function of $\mathsf{U}$, the result is usually called the *maximum power point* (MPP). The associated efficiency $\eta$ can then be plotted as a function of $E_g$. The numerically obtained solution is shown here for three representative cases in Figure 7.4 (a). The important result is the maximum of $\eta$ as a function of the bandgap $E_g$ in Figure 7.4 (a).

This result is most easily understood if one neglects the second term in Eq. (7.11) and looks at the $E_g$-dependence alone, which corresponds to $K_{out} = 0$ or $T_c = T_2 = 0$. The resulting efficiency is called the *ultimate efficiency*, $\eta_{ult}(E_g)$, which is represented by curve i) in Figure 7.4 (a). It can be expressed in the form

$$\eta_{ult}(E_g) = \frac{E_g \int_{E_g}^{\infty} f(E)E^2 \, dE}{\int_0^{\infty} f(E)E^3 \, dE}. \tag{7.12}$$

where $f$ is taken at $T = T_1$ and $\mu = 0$. The nominator counts the photons with $E > E_g$, and multilplies this number by the obtained energy per photon, $E_g$. The denominator counts the total energy of the incoming photons and is $E_g$-independent. Because $E_g$ appears in the upper integral as the lower integration boundary and also as a prefactor, the nominator goes to zero for $E_g \to 0$ and for $E_g \to \infty$, and must thus have a maximum in between these two limits. A numerical calculation yields for this maximum $E_g = 1.1$eV and $\eta = 44\%$. This result is universal, as the maximum efficiency does not depend on any parameter - it corresponds to the dimensionless value of $E_g/k_B T_1 = 2.2$. Figure 7.4 indicates that Si is the best conventional semiconductor material in what concerns the optimum bandgap for the ultimate efficiency.

If one takes into account a detailed balance for radiative recombination at a finite absorber temperature, the result depends on the ratio $K_{in}/K_{out}$. For $K_{in} \approx K_{out}$, there is only a small reduction of the efficiency, because the second integral in Eq. (7.11) is small due to $T_c = T_2 \ll T_1$. However, for $K_{in} \ll K_{out}$, there can be a stronger reduction of the efficiency. As a specific example we discuss the geometrical effect for a solar cell without concentration of the light by mirrors or lenses. As shown in Figure 7.4 (b), suppose the sun acts like a disk of area $\pi d^2/4$ with diameter $d = 1.4 \, 10^9$ m in a distance $L = 1.5 \, 10^{11}$ m from the earth. The solid angle from which the light is collected equals $\pi(d/L)^2/4 = 6.8 \, 10^{-5}$ sr. If the cell were a sphere, one would have to divide this by the full solid angle, $4\pi$, in order to obtain $K_{in}/K_{out}$. If it were a flat plate, one would have to calculate the radiation impinging on the surface $A$ on one cell side. Such geometrical considerations lead to an additional factor of about 0.1-1, and the final ratio is roughly $K_{in}/K_{out} \approx 10^{-5}$. This accuracy is sufficient if you take into account the roughness of all other approximations. So, have a look now at a typically resulting efficiency, given by curve iii) in Figure 7.4 (a). The efficiency maximum is in the 30 %-region at about $E_g \approx 1.25$ eV.

The electric power output, which is the interesting quantity for the receiver of the energy, can be calculated for a photovoltaic cell from Eq. (7.10). It turned out that the electrochemical potential, $\mu_c$, of the electron-hole pairs is related to the cell voltage $U = \mu_c/e$. You can understand the solar cell as a voltage source with a current−voltage characteristic

$$I(U) = I_{ph} - I_0(\exp(eU/k_B T) - 1) \tag{7.13}$$

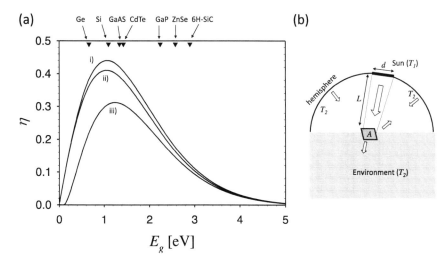

**Figure 7.4** (a) Efficiencies of a photo-diode as a function of the bandgap $E_g$. i) Ultimate efficiency ($K_{out} = 0$, $\eta_{max} = 44\%$ at $E_g = 1.1eV$) and detailed balance efficiencies (ii): $K_{out} = K_{in}$ ($\eta_{max} = 41\%$ at $E_g = 1.1eV$), and iii) $K_{in}/K_{out} = 2.18 \cdot 10^{-5}$ ($\eta_{max} = 31\%$ at $E_g = 1.25eV$). The triangles refer to the $E_g$-values of a few selected semiconductors. (b) Illustration for estimating $K_{in}/K_{out}$ associated with radiative power flow from the sun to the photovoltaic cell ($T_c = T_2$).

for $T = T_2$. The first contribution, $I_{ph}$, is the short-circuit current provided by the radiation. In presence of a load impedance, a voltage builds up, which counteracts via an additional negative current according to the Shockley diode characteristic. Therefore, the current-voltage characteristic of a pn-junction must be subtracted from $I_{ph}$, which is the second term in Eq. (7.13). The prefactor $I_0$ is the reverse saturation current. The current-voltage characteristics $I(\mathsf{U})$ is shown in Figure 7.5; the current $I(\mathsf{U})$ vanishes at the *open circuit voltage* value

$$\mathsf{U}_{oc} = \frac{k_B T_2}{e} \ln\left(1 + \frac{I_{ph}}{I_0}\right). \tag{7.14}$$

The relevant voltage range for Eq. (7.13) is thus $0 \leq \mathsf{U} \leq \mathsf{U}_{oc}$. The factor $I_0$ must be derived from semiconductor physics. We will not do this here but only give two remarks. First, for our simple assumptions in Section 7.1.2 we can get it from Eq. (7.11) in the following way. Because $E_g/k_B T_2$ is large, the 1 in the denominator of the Bose−Einstein function can be neglected such that a Boltzmann distribution $f \approx \exp\{(e\mathsf{U} - E_g)/k_B T_2\}$ follows. The factor $\exp(e\mathsf{U}/k_B T_2)$ may be put in front of $K_{out}$, such that $I_{ph}/I_0 = K \exp(E_g/k_B T_2)$. This approach to $I_0$ is, however, not very realistic as it includes only radiative recombination [DV00]. Secondly, a more general approach from semiconductor physics derives the reverse saturation current

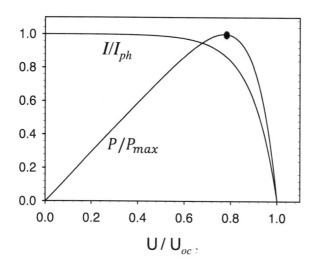

**Figure 7.5** Photodiode current-voltage characteristic and power-voltage characteristic with MPP for a pn-junction photo-cell. $U_{el}/U_{oc}$ can be interpreted as voltage efficiency, $\eta_U$.

from the dark generation rate, $G \propto \exp(-E_g/k_BT_2)$, in the active volume $lA$, by $I_0 = eGAl$; as mentioned earlier, $l$ is the width of the total diffusion region of the electrons and holes. In any case, if the Boltzmann factor with the gap energy predominates and $I_{ph} \gg I_0$, the open circuit voltage becomes

$$U_{oc} \approx \frac{E_g}{e}, \tag{7.15}$$

i.e., it is mainly determined by the bandgap.

Once $I(U)$ is given, the solar cell power is

$$P = UI(U), \tag{7.16}$$

which is shown in Figure 7.5. This curve can be seen as an efficiency-power relation, if $U/U_{oc}$ is interpreted as a voltage efficiency. It is obvious from the form of $I(U)$ that $P(U)$ has a maximum power point (MPP). To get the maximum power from the power source, the operation voltage must be tuned to the value where $d(UI(U))/dU = 0$, such that the load resistance $R = U/I$ satisfies the equation

$$R = -\frac{dU}{dI}. \tag{7.17}$$

This defines the $U_{MPP}$-value. Note that the maximum power is limited by $I_{ph}U_{oc}$; the ratio $P_{MPP}/I_{ph}U_{oc}$ is called *fill factor*, and serves as a figure of merit for photodiodes.

## 7.2 ELECTRICAL POWER CONVERSION DEVICES

Technologies for electrical power conversion by transformation between different voltage levels are at the heart of modern energy infrastructure. This section provides a brief insight in two examples, DC-DC (direct current) converters and AC-transformers. Further examples, like AC converters, can be discussed in an analogous manner.

### 7.2.1 DC-DC CONVERTER

A DC-DC converter is a device which transforms currents from one DC voltage level $\mathsf{U}_1$ to another DC level $\mathsf{U}_2$. We do not care here how such power electronics based−devices, which consist of semiconductor switches, diodes, and other electrical components, work in detail; here it is sufficient for you to imagine that it is something which just steps up or down the DC voltage. The ideal converter conserves the power. A general DC-DC converter can be modeled by the power balance equation

$$\mathsf{U}_1 I_1 = \mathsf{U}_2 I_2 + P_{loss}, \tag{7.18}$$

where the labels 1 and 2 refer to the input and output voltages and currents, respectively, and $P_{loss}$ models the converter losses. The efficiency-power relation is of course obtained from $P = U_2 I_2$ and $\eta = P/U_1 I_1$. The loss $P_{loss}$ is in general a function of $U_1$, $U_2$, $I_1$, $I_2$, and the switching frequency. It is often modeled as a second order polynomial in $P$. Loss contributions are, e.g., switching losses, conduction losses, losses from auxiliary equipment, capacitors and inductors. If, as a simple example, $P_{loss} = P_{NL} + RI_2^2$, consists of a no-load loss term $P_{NL}$ and a Joule type load-current dependent loss term, the efficiency-power relation becomes for prescribed (fixed) voltages $U_1$ and $U_2$

$$\eta(P) = \frac{P}{P + P_{loss}} = \frac{P}{P_{NL} + P + RP^2/U_2^2}, \tag{7.19}$$

which has an efficiency maximum at $\sqrt{P_{NL}U_2^2/R}$. A further discussion of the result is postponed to the next subsection where the same expression will reappear; it is just mentioned here that maximum DC-DC converter efficiencies can be larger than 95%; extreme efficiency power electronics can reach efficiencies > 99% [KKL+12].

### 7.2.2 TRANSFORMER

Let us continue with the efficiency of a transformer with losses [KK04]. You might correctly expect that they have quite similar efficiency-power relations as DC-DC converters. The transformation principle of the transformer will be postponed to Section 7.3.4 on impedance matching. The real transformer has magnetic and eddy current losses, $P_{NL}$, in the core (often iron), and conduction losses in the conductors (often copper). The so-called *iron losses*

(a)                                        (b)

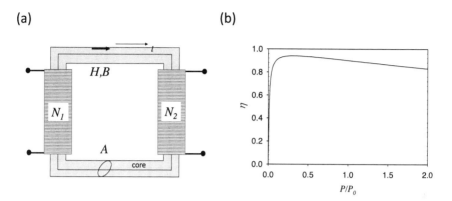

**Figure 7.6** (a) Illustration of an ideal transformer with primary and secondary winding numbers $N_1$ and $N_2$, length $l$ and cross-section area $A$ of the magnetic core. (b) Efficiency-power relation of a transformer or a converter with an efficiency maximum.

due to the oscillating magnetic field are present already without load and are thus responsible for no load losses. The *copper losses* scale typically with the square of the rms load current $\mid I_2 \mid$ and an effective resistance $R$, and are thus written in the form $R \mid I_2 \mid^2$. Because the rms voltages values $\mid U_1 \mid$ and $\mid U_2 \mid$ are usually fixed, and the power at the secondary (receiver) side is roughly given by $P = \mid U_2 \mid \mid I_2 \mid \cos(\theta_2)$, the efficiency can be written as

$$\eta = \frac{P}{P + P_{NL} + R \mid I_2 \mid^2}. \tag{7.20}$$

The power factor, $\cos(\theta_2)$, is the ratio of the real power and the apparent power. Using $\mid I_2 \mid = P/(\mid U_2 \mid \cos(\theta_2))$, one obtains in analogy to Eq. (7.19)

$$\eta(P) = \frac{P}{P_{NL} + P + P^2/P_0} \tag{7.21}$$

with $P_0 = (\mid U_2 \mid \cos(\theta_2))^2/R$. A typical efficiency-power curve, $\eta(P)$, is shown in Figure 7.6 (b). Suppose now that the power angle $\theta_2$ is more or less constant (we neglect any frequency dependence and assume a purely ohmic load). The efficiency maximum, $d\eta/dP = 0$, of the power transformer appears then at

$$P^* = \sqrt{P_{NL}P_0}, \tag{7.22}$$

where the copper losses equal the iron losses, $R \mid I_2 \mid^2 = P_{NL}$. The similarity with the results for the DC-DC converter is obvious. The maximum efficiency then has the value

$$\eta^* = \frac{1}{1 + 2\sqrt{P_{NL}/P_0}}. \tag{7.23}$$

The maximum efficiency-limit is governed by the no-load losses. Magnetic core optimization with respect to minimization of hysteresis and eddy current losses is thus an important task for transformer designers. High-power transformers can have quite high efficiencies in the 99% range, which means that the ratio $P_{NL}/P_0$ is rather small.

## 7.3 IMPEDANCE MATCHING

We have seen in several instances that there is an optimum load for which the power extracted from a source with internal losses becomes maximum. Already in the introduction it turned out that for an energy source with constant generalized force (e.g., an electric battery with open circuit voltage) $\mathsf{U}_0$ and internal resistance $R_0$, the maximum power occurs when the load resistance $R$ equals $R_0$ (this maximum-power law is sometimes called *Jacobi's law*). The maximum power value obtained from the parabolic efficiency-power relation is $P_{max} = \mathsf{U}_0^2/4R_0$ (see, e.g., Eqs. (1.2), (5.42), (6.15)). In practice, however, it requires some effort if you want to get the maximum power. Firstly, the impedance of the consumer is usually determined by the application device, and its resistance is mostly much larger than the preferably small internal resistance ($R_0 \ll R$) of the source. Secondly, the load resistance can vary with time, depending on the consumer needs. It can thus be desirable to couple the source and the load with a power transmission device such that the power is maximized by matching the effective impedance with the source. Examples are illustrated in Figure 7.7. An important application is *maximum power point tracking* (MPPT) in photovoltaic devices. Since this is a direct-current (DC) issue, we will briefly discuss impedance matching with a DC-DC converter in the next subsection, before MPPT is explained. In case of alternating currents (AC), the load can be matched with the help of transformers or appropriate LC-circuits, as will be illustrated in subsequent sections. It should be clear that although we will mainly look at electrical examples, the concept of impedance matching is more general and can be applied to other energy-flow types.

Before we start, a short remark on the maximum power law in Section 1.2 is added concerning finite frequencies and complex impedance. Suppose, you neglect leakage, and have given the impedances $Z_0 = R_0 + i\mathcal{X}_0$ for the source and $Z = R + i\mathcal{X}$ for the load, with reactances $\mathcal{X}_0$ and $\mathcal{X}$. What is the complex load impedance $Z$ that maximizes the power for a harmonic voltage, $\sqrt{2}U_0\cos(\omega t)$? The total power $P_{tot}$ is the time average of $I(t)U_0(t)$. In frequency space, one has for the voltage $U_0\exp(i\omega t)/\sqrt{2}$, a current $I_0\exp(i\omega t)/\sqrt{2}$ with $I_0 = U_0/Z_{tot}$, where $Z_{tot} = Z_0 + Z$, and thus $I(t) = I_0\exp(i\omega t)/\sqrt{2} + I_0^*\exp(-i\omega t)/\sqrt{2}$, where the $*$ indicates the complex conjugate. With this, the time average of $I(t)U_0(t)$ becomes $(1/Z_{tot} + 1/Z_{tot}^*)U_0^2/2$,

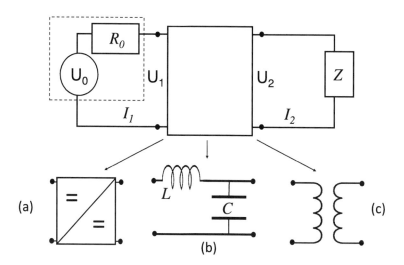

**Figure 7.7** Impedance matching problem: a power source (dashed box) with open circuit voltage $U_0$ and internal resistance $R_0$ feeds a load with impedance $Z$. In order to maximize the transferred power, case-dependent electric elements can be inserted, for instance (a) DC-DC-converters (Section 7.3.1), (b) LC-circuits (Section 7.3.3), or (c) transformers (Section 7.3.4).

which gives

$$P_{tot} = \frac{R + R_0}{(R + R_0)^2 + (\mathcal{X} + \mathcal{X}_0)^2} U_0^2 = P + P_s, \qquad (7.24)$$

which is the sum of the two losses at the load and the source impedances, $Z$ and $Z_0$, respectively. The power $P$ at the load $Z$ is the part of Eq. (7.24) which contains the $R$ in the nominator. To find the maximum of $P$, we can now vary $\mathcal{X}$ and $R$. Obviously, $\mathcal{X}$ appears only in the denominator, which is minimized if $\mathcal{X} = -\mathcal{X}_0$. Once this is done, everything goes as for the DC case: $R = R_0$. We may thus conclude, that the power is maximum for $Z = Z_0^*$, i.e., $Z$ is the complex conjugate of $Z_0$. This means compensation of the source reactance.

## 7.3.1 IMPEDANCE MATCHING WITH A DC-DC CONVERTER

Suppose that the circuit in Figure 7.7 is a DC circuit with real impedance $Z = R \neq R_0$. The DC-DC converter with voltage ratio $\chi = U_1/U_2$ and constant converter losses $P_{loss}$ can then be directly used for maximizing the power. From $U_0 = R_0 I_1 + U_1$, Eq. (7.18), and $P = U_2^2/R$, one can solve for $U_2$ as a function of $\chi$, and write down $P(\chi)$ and $\eta(\chi) = P/(I_1 U_0)$. For the most simple case $P_{loss} = 0$, one gets

$$P(\chi) = \frac{\chi^2}{(R_0/R + \chi^2)^2} \frac{\mathsf{U}_0^2}{R}, \tag{7.25}$$

and

$$\eta(\chi) = \frac{\chi^2}{R_0/R + \chi^2}. \tag{7.26}$$

Obviously, a power maximum occurs at $\chi = \sqrt{R_0/R}$ with $\eta = 1/2$. We will find a similar result for the AC-transformer in Section 7.3.4. You can draw the curves $P(\chi)$ and $\eta(\chi)$, and derive the $P(\eta)$-relation by elimination of $\chi$ from the Eqs. (7.25) and (7.26). It leads again to Eq. (1.2). To plot the curve for finite converter losses $P_{loss}$ is also straightforward, it looks as you might expect from Figure 1.1; in particular the curve is closed and also exhibits an efficiency maximum.

### 7.3.2   MAXIMUM POWER POINT TRACKING

We return to the solar cell of Section 7.1.2. Because $I(\mathsf{U})$ changes with varying irradiance $I_{ph}$ and temperature $T_2$, the maximum power point $\mathsf{U}_{MPP}$ fluctuates in time. If you want to stay at maximum power you must follow the MPP, and this is what is called *maximum power point tracking* (MPPT). In the following we suppose that the changes are slow enough for assuming a quasi steady-state. In order to track the MPP, one needs two ingredients then. The first is an electric device which allows measuring and controlling the voltage value $\mathsf{U}$. The second is an algorithm, which pulls the voltage value $\mathsf{U}$ towards $\mathsf{U}_{MPP}$. You can imagine that the engineers developed various control algorithms of different complexity which converge towards $\mathsf{U}_{MPP}$ as a fix-point. The simplest one is maybe the *fractional open circuit voltage*, which just puts $\mathsf{U} = k\mathsf{U}_{oc}$ heuristically (see Eq. (7.14)) with an appropriate factor $k$ ($0 < k < 1$). This usually works because the fraction $\mathsf{U}_{MPP}/\mathsf{U}_{oc}$ is often roughly constant in a large range of $I_{ph}$- and $T$-values. We will not go into further details on MPPT algorithms.

As an application example for MPPT, let us combine the photodiode characteristics with a loss-free ($P_{loss} = 0$) DC-DC converter and consider a photovoltaic cell (subscript 1) which is used for charging up an ideal battery (subscript 2). The ideal battery requires constant voltage $\mathsf{U}_0$. Without DC-DC converter, the battery is charged with current $I_{ch}^0 = I(\mathsf{U}_0)$ from Eq. (7.13). With a converter that transforms from $\mathsf{U}_1 = \mathsf{U}_{MPP}$ to $\mathsf{U}_2 = \mathsf{U}_0$, the current on the battery side becomes

$$I_{ch}^{MPP} = \frac{\mathsf{U}_{MPP}}{\mathsf{U}_0} I_{MPP}. \tag{7.27}$$

Obviously, $I_{ch}^{MPP}/I_{ch}^0 = P_{max}/P(\mathsf{U}_0) \geq 1$, i.e., the DC-DC converter leads in this way to the fastest charging of the given battery with the given photovoltaic power source.

### 7.3.3  IMPEDANCE MATCHING WITH AN LC CIRCUIT

Many applications are based on finite frequencies or alternating currents (AC) rather than on steady state or direct currents (DC). We show now that with an appropriate choice of capacitance $(C)$ and inductance $(L)$ values in the circuit in Figure 7.7 (b), the power extracted by a load resistance $R$ from the AC source (e.g., a generator) with voltage $\mathsf{U}_0$ and internal resistance $R_0$ can be maximized. It will turn out that this specific circuit requires that the load and source impedances satisfy a certain condition, but that is no problem because otherwise a similar circuit can be used. The method is common in signal transmission technology, but we consider it here as a general way to match the load for power maximization in frequency dependent systems. Sophisticated theoretical methods exist to solve the general problem, but for the simple case here we can do the calculation directly. We go to the frequency space and consider complex quantities, like for the voltage $\mathsf{U}_0(\exp(i\omega t)+\exp(-i\omega t))/\sqrt{2}$ with (cosine) amplitude $\sqrt{2}\mathsf{U}_0$, and thus effective (or rms-) voltage $U_0$. The complex impedance $\mathsf{U}/I$ of a resistor is real, $R$, and for a capacitor and an inductor it is imaginary, $1/i\omega C$ and $i\omega L$, respectively. According to the circuit in Figure 7.7 (b), we have

$$\mathsf{U}_0 = R_0 I_1 + i\omega L I_1 + R I_2, \tag{7.28}$$

with capacitor voltage

$$R I_2 = \frac{1}{i\omega C}(I_1 - I_2). \tag{7.29}$$

The latter equation gives $I_1 = (1 + i\omega C R)I_2$, hence

$$\mathsf{U}_0 = ((R_0 + i\omega L)(1 + i\omega C R) + R)\, I_2, \tag{7.30}$$

such that the power dissipated at the load $R$ becomes

$$P = R\,|\,I_2\,|^2 = \frac{1}{(1 + R_0/R - \omega^2 LC)^2 + \omega^2(R_0 C + L/R)^2} \cdot \frac{\mathsf{U}_0^2}{R}. \tag{7.31}$$

The contour lines of $P(C, L)$ are shown in Figure 7.8. The power can now be maximized by solving the equations $dP/dC = 0$ and $dP/dL = 0$, which read

$$-2(1 + R_0/R - \omega^2 LC)\omega^2 L + 2\omega^2(R_0 C + L/R)R_0 = 0 \tag{7.32}$$
$$-2(1 + R_0/R - \omega^2 LC)\omega^2 C + 2\omega^2(R_0 C + L/R)/R = 0. \tag{7.33}$$

By comparing the two equations one concludes

$$C R_0 R = L. \tag{7.34}$$

Substitution of $C = L/R_0 R$ in Eq. (7.32) leads to the optimum inductance and capacitance values,

$$\omega L = \sqrt{R_0(R - R_0)}, \tag{7.35}$$

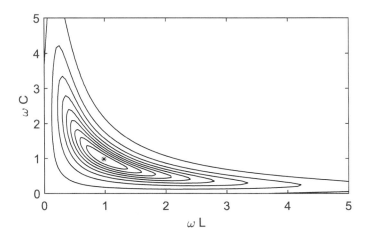

**Figure 7.8** Coutour lines of $P(C, L)$ from Eq. (7.31) for the values $R_0 = 0.2\Omega$ and $R = 5\Omega$, $f = 50$ Hz (more realistic $R/R_0$-ratios lead to a less visible maximum, because the contours are strongly squeezed to the hyperbola $w^2 LC \approx 1$.)

$$\omega C = \sqrt{R^{-1}(R_0^{-1} - R^{-1})}. \tag{7.36}$$

Putting these expressions into Eq. (7.31) leads to the well-known result $P_{max} = \mathsf{U}_0^2/4R_0$. Note that $\omega^2 LC = 1 - R_0/R$. The example circuit considered works for $R > R_0$; otherwise, other circuits can be used, e.g., by a rearrangement of the $L$ and $C$. A further interesting example concerns the impedance of a long power transmission line composed of a long series of the above considered $L$-$C$-circuits. One can show that its impedance is the geometric mean of the capacitive and inductive impedances, $\sqrt{L/C}$, and its matching is certainly relevant for power transmission. Nevertheless, we will not go into the details; the goal here was only to show that with an appropriate passive circuit, one can get the power maximum from the AC power source. Note that the values of the circuit parameters depend on frequency.

### 7.3.4 IMPEDANCE MATCHING WITH A TRANSFORMER

As a last example, we discuss how impedance matching works with a (lossless) AC-transformer, as an analogy to the DC-DC converter. We will again consider purely ohmic impedances of source and load for a simple illustration of the concept (from Eq. (7.24) we learned that, in the presence of reactances, they must be compensated in order to obtain maximum power). We take the time and do the calculations in some detail. First, we recall the basic equations for the ideal loss-free transformer, as symbolized in Figure 7.7 (c). It transforms the voltages and currents between primary and secondary sides by a set of inductances $L_{jk}$, such that the voltages can be written as

$$\mathsf{U}_1 = i\omega L_{11}I_1 + i\omega L_{12}I_2 \tag{7.37}$$
$$\mathsf{U}_2 = i\omega L_{21}I_1 + i\omega L_{22}I_2. \tag{7.38}$$

To understand this, consider Figure 7.6 (a). The closed, soft-ferromagnetic core (i.e., without air gap, negligible hysteresis, and far below magnetic saturation) of the ideal transformer is supposed to have a very high magnetic permeability $\mu_r$ such that all magnetic field lines, $(H)$, and flux density lines, $(B = \mu_0\mu_r H)$, are concentrated inside the core without any loss of the magnetic flux $BA$, with $A$ being the cross section area of the core. Ampere's law in its integral form,

$$\oint H_k \, dl = N_k I_k, \tag{7.39}$$

where the integration path goes along the core of length $l$, implies that the current $I_k$ ($k = 1$ and 2) in the coil $k$ with $N_k$ windings leads to a H-field contribution obtained from $lH_k = N_k I_k$. The index $k$ of $H_k$ indicates here by which coil it is generated; the total $H$ field is the sum of both contributions. The magnetic flux through the core is then $B = \mu_r\mu_0(H_1 + H_2)$. Hence, the total flux through coil $j$ with $N_j$ windings, becomes $N_j BA$. The voltage $\mathsf{U}_j$ induced in this coil is given by the time derivative of this flux, which means in frequency space multiplication by $i\omega$. This leads immediately to the Eqs. (7.37) and (7.38), with the inductances

$$L_{jk} = \frac{\mu_0\mu_r A}{l} N_j N_k. \tag{7.40}$$

Obviously, one has $\mathsf{U}_k = i\nu N_k(I_1 N_1 + I_2 N_2)$ with $\nu = \omega\mu_0\mu_r A/l$, which strictly implies

$$\frac{\mathsf{U}_1}{\mathsf{U}_2} = \frac{N_1}{N_2}. \tag{7.41}$$

*Strictly* means that it holds for any loading of our ideal transformer. Now, we put a load with real impedance $Z = R$ to the secondary side, such that $\mathsf{U}_2 + RI_2 = 0$. Elimination of $\mathsf{U}_2$ in Eq. (7.38) gives

$$I_1 = -\frac{R + i\nu N_2^2}{i\nu N_1 N_2} I_2. \tag{7.42}$$

$I_1/I_2 = N_2/N_1$ holds obviously for no-load conditions, $R = 0$. Having a power source on the primary side, with $\mathsf{U}_1 + R_0 I_1 = \mathsf{U}_0$ one can solve Eq. (7.37) for $\mathsf{U}_0$:

$$\mathsf{U}_0 = -\frac{I_2}{i\nu N_1 N_2}\left(R_0 R + i\nu(R_0 N_2^2 + RN_1^2)\right), \tag{7.43}$$

and eventually write the load power as (assuming real $\mathsf{U}_0$)

$$P = R\,|\,I_2\,|^2 = \frac{R\nu^2 N_1^2 N_2^2 \mathsf{U}_0^2}{R_0^2 R^2 + \nu^2(R_0 N_2^2 + RN_1^2)^2}. \tag{7.44}$$

If you try to find a maximum of $P(N_1, N_2)$ by solving $dP/dN_1 = dP/dN_2 = 0$ for finite $N_k$, you will fail, because the true maximum is at $N_k \to \infty$ with a given ratio $N_1/N_2$. Infinite ($\infty$) means here, that the $N_k$ values are so large that the first term in the denominator of Eq. (7.44) can be neglected. In other words, the transformer inductive reactances should be much larger than the resistances. Then, one can write the power as

$$P = \frac{R \mathsf{U}_0^2 N_1^2 / N_2^2}{(R_0 + R N_1^2 / N_2^2)^2}, \tag{7.45}$$

which is equivalent to Eq. (7.25) with $\chi = N_1/N_2$, and leads thus to

$$\frac{N_1}{N_2} = \sqrt{\frac{R_0}{R}} \tag{7.46}$$

with maximum power $P_{max} = \mathsf{U}_0^2 / 4R_0$. No surprise at all. For $R_0 < R$, a step-up transformer ($\mathsf{U}_1 < \mathsf{U}_2$) is required, and for $R_0 > R$ of course a step-down transformer is needed.

## 7.4  ELECTRO-MECHANICAL ENERGY CONVERSION: MOTORS

Without exaggeration one may state that electrical generators and motors are among the most important devices of modern civilization. There is a huge power range where electrical motors and generators are operating, from nW to GW! Particularly for large apparatuses, be it for industrial or transportation applications, high efficiency is crucial for saving energy costs. Of course, efficiencies $\eta$ of different devices increase as a function of the power rating $P$; very large motors, for example, can have efficiencies up to 99%! Nevertheless, transformers can reach higher efficiency limits than motors with comparable power ratings, as the latter have moving parts with additional friction losses. The main purpose of this section is to show how electro-mechanical energy conversion devices can be described by efficiency-power relations for steady-state operation. Real designs of motors and generators are adapted to the specific applications where they are used, which leads to a large variety of different generator and motor types. Generators refer to the same devices with just an exchange of the device being a power source instead of a power sink. We will thus mainly focus on motors.

The diversity of real motor types is larger than for generators, and if you are interested in a detailed classification of the electrical motors according to the different working principles you should dig into the huge body of literature on electrical engines and related topics like motor drives (see, e.g., [HD13]). In the following, we will focus on electrical motors where the Lorentz force and the induction law play the main role, while other types like piezo-electric motors will not be discussed despite their importance for many applications. Furthermore, not all motors are made to provide rotational energy. For instance the task of servo motors is to control acceleration, velocity, and position of an object. One usually distinguishes between motors where the electric current (or voltage) is unipolar (direct current, DC) and where it is alternating (AC). Most of today's electro-mechanical power conversion devices transform an electric AC power, with harmonic time-dependent current $I(t)$ and voltage $U(t)$, into mechanical rotational motion, with angular frequency $\omega$ of the rotating part (called the *rotor* or *armature*) and torque $M$ between the rotor and the stationary part (called *stator*). If the electrical AC-oscillation and the mechanical rotation are synchronized in steady-state operation, the AC machine is called *synchronous*, otherwise *asynchronous*. From a historical perspective, the DC motor was the dominant technology before the breakthrough of electronics and power electronics in particular. In the second half of the last century the picture reversed due to the development of power-electronic motor drives for AC motors [HD13]. Because we emphasize the general *concept*, it is sufficient to discuss a simple prototype DC-motor. Its convenient property is the time-constancy of the power conversion. A single phase AC-motor would have a sinusoidal time dependence of power and force or torque. However, for three electric phases, which is the usual case for industrial AC motors, power is also constant in time, such that our simple model will be sufficient to understand the concept.

### 7.4.1   TORQUE-FREQUENCY RELATION AND STABILITY

Below we will restrict the considerations to steady-state conditions needed for the derivation of the efficiency-power relations. Nevertheless, a remark on the transient motor dynamic is important, because it is related to the *stability* of the steadily rotating state. If the rotor's constant moment of inertia is $\mathcal{J}$, the equation of motion for the angular frequency $\omega$ reads

$$\mathcal{J}\frac{d\omega}{dt} = M(\omega) - M_e, \qquad (7.47)$$

where $M_e$ is the external load torque (its sign is negative since it acts to decelerate the angular velocity). Stationary operation refers to the steady-state frequency value $\omega_0$, which is the solution of $M(\omega) = M_e$. For $M(\omega) \neq M_e$ the rotational motion is either accelerated or decelerated, depending on whether $M(\omega) > M_e$ or $M(\omega) < M_e$, respectively. This immediately implies that the frequency-torque characteristics $M(\omega)$ must have negative slope in order for the steady state to be stable, since otherwise a small frequency deviation $\Delta\omega = \omega - \omega_0$ from the steady-state frequency $\omega_0$ would grow instead of decay. You can see this if you write Eq. (7.47) for $\Delta\omega$ up to the first order of a Taylor expansion,

$$\mathcal{J}\frac{d\Delta\omega}{dt} = \frac{dM}{d\omega}\Delta\omega, \qquad (7.48)$$

which has the time-dependent solution $\Delta\omega(t) \propto \exp(\mathcal{J}^{-1}(dM/d\omega)t)$. We will derive below $M(\omega)$ curves, and will find that they do satisfy $dM/d\omega < 0$.

Let us have a look at the simple illustration of a linear DC motor in Figure

(a)                                    (b)

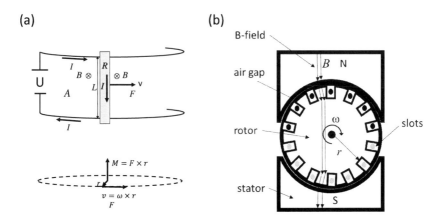

**Figure 7.9**  (a) Illustration of a linear DC motor consisting of a current-conducting rod gliding along two rails in a perpendicular $B$-field. (Bottom: in the rotating machine, force and velocity are analogous to the torque and angular frequency, respectively.) (b) In the real rotating machine, the current changes direction when passing from the upper (black cross sections) to the lower half (white cross sections) by a commutation mechanism. In order to minimize the magnetic losses due to the air gap between the stator and rotor, the conductors are embedded in slots.

7.9 (a). It consists of a rod of length L with resistance $R$, which is able to slide without friction on two parallel rails with negligible resistance. The rod plays the role of the rotor and the rails of the stator. Instead of $\omega$ and torque $M$, which are the relevant quantities in motors with rotational motion, you may also consider here for linear motors the velocity $v$ and force $F$ instead, by replacing $F$ by $M$ and $v$ by $\omega$, because the corresponding quantities are related by constant factors. $M = F \times r$ and $v = \omega \times r$, with radius $r$ of the rotor, as is indicated at the bottom of the figure. As the motor is characterized by the functional relation between $M$ and $\omega$ (or $F$ and $v$), you know immediately the mechanical power $P_m(\omega)$ (or $P_m(v)$)

$$P_m(\omega) = F(v)v = M(\omega)\omega, \tag{7.49}$$

once you know the function $M(\omega)$ (or $F(v)$). Suppose a battery with constant voltage U is connected to the rails such that a current $I$ is flowing through the rod. Additionally, there is a perpendicular magnetic flux density $B$. One distinction between different motor types lies in the way to create the $B$-field. We mention three different cases. In the first case, the $B$-field is created by a permanent magnet and is just constant. In the second case, it is created by a magnetic coil which is connected in parallel to the rod; this is the so-called *shunt* configuration. In our case, $B$ would also be constant because the current through the coil, given by the voltage $\mathsf{U}$ and the coil resistance, is so; of course this current would add to the electric loss depending on the ohmic resistance of the coil. In the third case the magnetic coil is connected in series to the rod (i.e., the rotor coil), which leads to a $B$-field which is proportional to the rotor current. To make the story not too long, we discuss the first and last cases only.

## 7.4.2  EFFICIENCY-POWER RELATIONS OF DC MOTORS

We consider first the most simple permanent magnet motor. You only need to remember two basic items from your introductory courses in electromagnetism: the *Lorenz force* and the *induction law*. The Lorenz force

$$F = IBL \tag{7.50}$$

acts on the straight conductor of length $L$ which carries a current $I$ in a perpendicular $B$-field. This force accelerates the rod along the rails. The motion of the rod with velocity $v$ leads to an induced voltage according to the induction law

$$\mathsf{U}_{ind} = -\frac{d\Phi}{dt}, \tag{7.51}$$

where $\Phi = BA$ is the magnetic flux through the area $A$ enclosed by the rails, the rod, and the battery. Because $B$ and $L$ are constant, $d\Phi/dt = BdA/dt = BLv = BLr\omega$, hence

$$\mathsf{U}_{ind} = -BLr\omega. \tag{7.52}$$

The current-voltage relation of the circuit is thus

$$\mathsf{U} = RI + BLr\omega. \tag{7.53}$$

Elimination of $I$ with the help of Eq. (7.50) gives the relation between frequency and torque, which we write in the form

$$M = \hat{M}\left(1 - \frac{\omega}{\hat{\omega}}\right), \tag{7.54}$$

where

$$\hat{M} = \frac{\mathsf{U}BLr}{R}, \tag{7.55}$$

is the *driving torque*, i.e. the torque at zero velocity and thus also the torque at motor start-up, and

$$\hat{\omega} = \frac{\mathsf{U}}{BLr}, \tag{7.56}$$

is the *idle frequency*, corresponding to the velocity at zero torque and thus at no-load conditions. It is immediately obvious from Eq. (7.49) that $P_m = 0$ at the two points $\omega = 0$ and $\omega = \hat{\omega}$, and that there must be a maximum of $P_m$ in between. For the further procedure it is convenient to introduce

$$P_0 = \hat{M}\hat{\omega} = \frac{U^2}{R},$$
(7.57)

such that the mechanical power becomes

$$P_m = P_0 \left(1 - \frac{\omega}{\hat{\omega}}\right) \frac{\omega}{\hat{\omega}}.$$
(7.58)

The total power provided by the electrical power source is

$$P_{el} = UI = P_0 \left(1 - \frac{\omega}{\hat{\omega}}\right).$$
(7.59)

The efficiency of the motor is then given by the ratio of electrical to mechanical power

$$\eta = \frac{P_m}{P_{el}} = \frac{\omega}{\hat{\omega}}.$$
(7.60)

One can thus express the motor power as a function of the efficiency

$$P_m(\eta) = P_0\eta(1 - \eta),$$
(7.61)

which has a maximum at $\eta^* = 1/2$ with maximum power $P_0/4$. You know this result from earlier, e.g., Eq. (1.2), so you might have predicted it. Indeed, our system is nothing than a battery power-source with internal series resistance, connected to a load (here a motor).

Of course, the restriction of the losses to conductor losses is much to restrictive for a general (AC, rotating, ....) motor. There are plenty of other loss channels, e.g., related to mechanical friction (bearing), eddy currents, magnetic hysteresis, etc.. For real machines they must be considered when deriving the relation between $P_m$ and $\eta$. The general procedure, however, remains the same.

In the following, we discuss one example which goes beyond the $M(\omega)$-decrease with constant slope, namely, the above-mentioned third case where the $B$-field is produced by the stator coil which is connected in series to the rotor (our rod). The difference to the previous case is that $B$ is not fixed but a function of the current,

$$B = KI,$$
(7.62)

with a constant $K$ that depends on the specific case. You certainly know that the ratio of the magnetic flux, $BA$ and the current $I$ is equal to the inductance, so $K$ is related to the inductance and is thus a geometrical characteristic quantity. We combine Eqs. (7.53) and (7.62) and obtain

$$I = \frac{U}{R + LKr\omega},$$
(7.63)

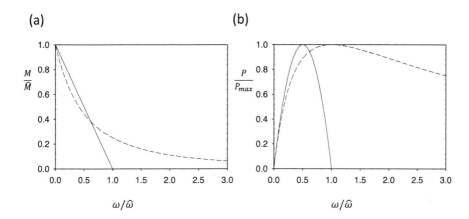

**Figure 7.10**  (a) Frequency-torque relations for the motors with permanent magnet (solid curve) and series connected rotor and stator field (dashed). (b) Frequency-power relations for the respective cases in (a). The efficiency-power relation is given by Eq. (7.61).

which leads to the frequency-torque relation

$$M = rKLI^2 = \frac{U^2}{R^2}\frac{KLr}{(1+\omega/\hat{\omega})^2} = \frac{\hat{M}}{(1+\omega/\hat{\omega})^2},\qquad(7.64)$$

with characteristic frequency

$$\hat{\omega} = \frac{R}{LKr},\qquad(7.65)$$

and initial torque

$$\hat{M} = \frac{rKLU^2}{R^2}.\qquad(7.66)$$

It is now straightforward to determine the efficiency $\eta = P_m/P_{el} = M\omega/UI$, which yields

$$\eta = \frac{\omega}{\omega + \hat{\omega}}.\qquad(7.67)$$

If you solve for $\omega(\eta)$ and replace this in $P_m = M\omega$, you find that the efficiency-power relation is (of course!) again given by Eq. (7.61).

Let us compare the two different cases. The two frequency-torque relations are shown in Figure 7.10. Obviously, while in the first case, Eq. (7.59) led to the finite idle-frequency value (7.56) at no-load conditions, now the frequency can become arbitrarily large, i.e., the idle frequency diverges. In both cases the maximum motor speed is independent of the resistance $R$. This is to be expected because the two states correspond to the efficiency $\eta = 1$, where losses are absent. Indeed, the current vanishes, $I = 0$, in the ideal idle motion. In reality the speed is of course limited by other loss factors like friction, etc., and the motor might need a protection mechanism against damage due to too high motor speed at no-load.

## 7.5   FLUID-FLOW POWER

We transition now into the last topic of this chapter, namely, energy conversion from a flowing fluid into mechanical rotational energy of a turbine wheel. The mechanical power can then further be transformed in electrical or other energy forms, but we will focus here on the mechanical power output.

There are several ways to classify the different turbine technologies. First, one can distinguish between *free* and *ducted* fluid-flow energy conversion. As will become clear below, free-flow power-conversion machines, like wind wheels, have a fundamental limit for the maximum power fraction that can be extracted from the kinetic fluid energy, due to the eluding reaction of the flow to the obstacle. For ducted flow, where the fluid cannot laterally escape from the active turbine area, e.g., because of a casing, a pipe, or a concentrated liquid jet, the ideal conversion efficiency can in principle become close to 100%. Large hydro-power plants can reach electrical power efficiencies larger than (95%)!

A further refinement of a physical classification is connected to the interaction of the fluid with the wheel or turbine. Some basic examples are illustrated in Figure 7.11. In the following subsections we will determine the efficiency for *lift force* and *drag force* based free-flow turbines. In a nutshell, the lift force is perpendicular to, while the drag force is in the direction of the fluid flow velocity, as is depicted in Figure 7.11 (a). Real turbines usually exploit both force components. Another differentiator for a classification refers to the position and alignment of the wheel axis, i.e., whether it lies *vertical* or *horizontal*, and - for water wheels - if the horizontal axis is *below* or *above* water level.

An important classification of turbines with ducted fluid inflow separates *reaction turbines* from *impulse turbines*, which utilize, respectively, the reaction force (or pressure) and the kinetic impact energy of the flow. In impulse turbines, the flow is usually concentrated with a nozzle into a jet that impinges the runner vanes. As an example, we will discuss the *Pelton turbine* (cf. Figure 7.11 (d)) where the kinetic energy is removed from the jet by the blades, and no pressure casement is necessary. Hydraulic reaction turbines, on the other hand, make additionally use of the pressure difference across the turbine. They run, in part, according to the reaction force of the fluid outflow in a way similar to a garden sprinkler which rotates due to the *recoil force* in response to the ejected water. Since general fluid flow (gas, liquid, steam) contains all three energy forms (kinetic, pressure, and thermal), general turbines can be characterized by their *degree of reaction*, which is defined by the fraction of enthalpy (pressure/thermal power) that contributes to the to-

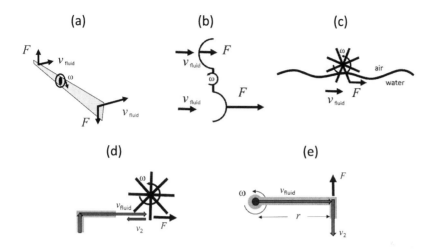

**Figure 7.11** Different types of fluid-flow power conversion principles. Free flow: lift force (a), drag force (b) and (c); ducted flow: impulse (d) and reaction (e) turbines.

tal power, which contains additionally the kinetic power. Steam turbines can have different stages with different degrees of reaction.

The origin of hydro-energy lies of course in the specific gravitational energy density, $\rho g h$, of the upstream reservoir. An arrangement of different water-turbine technologies is often performed according to their application range with respect to volume flow, $\dot{V}$ (m$^3$/s) and drop height difference, $h$ (water head). Whether the pressure drop is small (e.g., river hydraulic power stations) or large (e.g., pumped hydro power) partly decides the optimum turbine type. An introduction and discussion of the various turbine types, like *Kaplan* turbines for low heights (up to several tens of meters), *Francis* turbines for average heights (tens to several hundreds of meters), and *Pelton* turbines for even higher heads go beyond our purpose. Similarly to electrical generators, which we did not explicitly discuss because they are reversely operated motors, we will not talk about propellers, fans, pumps, and the same here, but note that they can be understood as reversely operated turbines. In the following, a few prototype examples will be discussed for illustration of the working principles and the different turbine efficiencies: lift and drag force free flow, and impulse and reaction force ducted flow turbines. The section closes with a brief remark on efficiency-power relations for real turbines. The special cases that will be considered cover only a small part of the general topic of turbines. The consideration of thermal energy is completely omitted, and even some issues relevant for the examples discussed will not be introduced, such as the *Euler's turbine equation* and *pump equation*. If you want to explore the subject in depth, see, e.g., Refs. [BSJB01, Qua08, CC13] and other appropriate engineering textbooks.

### 7.5.1 LIFT-FORCE BASED FREE FLUID-FLOW POWER

One can understand the *atmospheric wind generator* driven by solar irradiation in the framework of endoreversible thermodynamics [DV08]. However, we simply suppose here that the wind is constantly blowing and estimate the energy current-density. With mass-density $\rho \approx 1.2$ kg/m$^3$ and velocity $v_1$, the kinetic energy density is $w_{kin} = \rho v_1^2/2$. We neglect density variations and compressibility effects. The kinetic energy current-density is

$$j_w = w_{kin} v_1 = \frac{1}{2} \rho v_1^3. \tag{7.68}$$

Typical wind velocities $v$ range from 0.1 m/s (practically windless) over 5 m/s (weak wind, ca. 75 W/m$^2$) up to 25 m/s (storm, 10 kW/m$^2$) and even > 30 m/s (hurricanes). The power flux through an area $A$ is

$$\dot{W}_1 = \frac{1}{2} \rho A v_1^3. \tag{7.69}$$

Diameters of wind wheels nowadays reach values on the order of 100 m, which leads to roughly 1 MW for a wind speed of 6 m/s.

The energy current-density carried by the resulting wind fluctuates strongly in space and time. Therefore, for practical purposes of wind energy harvesting, the statistical distribution function of the wind velocity is an important characteristic figure of a geographic location, where wind power plants are to be installed.

What is the maximum power that can be extracted from wind for a given wind velocity? The well-known off- and on-shore horizontal-axis wind-power turbines are mainly based on lift-force driven blades. To model this we invoke the three main equations of fluid mechanics: the conservation laws of *mass*, *momentum*, and *energy*. In Figure 7.12 (a) the flow lines of the wind in the vicinity of the wind wheel are shown by arrows before and after the rotor. Only those flow lines which just touch the tip of the blades are fully drawn; they enclose the flow channel relevant for the conversion of wind energy to the rotation energy of the wind wheel. This picture is of course an idealization. For instance, it corresponds to the limit of infinitely many blades, while the decelerated air flow can still pass through the area $A$ without friction power loss (the only power lost by the wind is what is transferred to the rotor for harvesting). In any case, it is clear that due to the extraction of kinetic energy, the wind velocity $v_2$ behind the rotor is smaller than $v_1$. This is also shown in Figure 7.12 (a). Behind the rotor, the downstream area $A_2$ is then larger than $A_1$. More specifically, the mass continuity equation requires a constant mass current in the flow channel

$$J_m = \rho_1 v_1 A_1 = \rho v A = \rho_2 v_2 A_2. \tag{7.70}$$

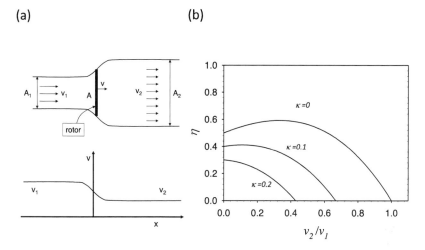

**Figure 7.12** (a) Illustration of a wind turbine with inflowing air (index 1, up-stream) and outflowing air (index 2, downstream); underneath: the velocity profile along the axis. (b) Efficiency as a function of the velocity ratio for different values of the loss parameter $\kappa$ with Newtonian friction $n = 2$. The optimum power of the ideal ($\kappa = 0$) wind turbine is obtained when $v_2 = v_1/3$ (Betz's law).

Because momentum and energy are transferred to the wind wheel, the associated balance equations for the momentum density, $\rho v$, and kinetic energy density, $w_{kin}$, contain some sink terms associated with transmitted momentum and converted power. The work done by a velocity change $dv$ in the volume $A\,\Delta x$ is given by $d\Delta W = \Delta m v dv = \rho A \Delta x\, v\, dv = J_m \Delta x\, dv$. Since $J_m$ is constant, integration gives $\Delta W = \Delta x J_m (v_2 - v_1)$. The force is thus $-\Delta W/\Delta x = J_m(v_1 - v_2) = \rho_1 A v_1(v_1 - v_2)$. Alternatively, one can also proceed via the difference in the momentum flows, and obtain for the total force

$$F_m + bv^n = \rho_1 v_1^2 A_1 - \rho_2 v_2^2 A_2 = \rho_1 v_1^2 A_1 \left(1 - \frac{v_2}{v_1}\right), \qquad (7.71)$$

where Eq. (7.70) was used for the last equality. The total force on the left-hand side is the sum of the mechanical force on the rotor for momentum and energy transfer, and an additional friction force modeled heuristically with a power law $v^n$. This friction loss model is culpably oversimplifying the situation. In the extensive wind energy literature the different loss types, like blade tip losses and wake losses, have been investigated thoroughly, but this goes again beyond our purpose of simply illustrating the general concept. So we suppose the above phenomenological friction function, just because it models in a simple way the losses. The difference in the energy flows is the sum of mechanical power plus friction-loss power

$$\dot{W}_m + b v^{n+1} = \frac{1}{2} \rho_1 v_1^3 A_1 - \frac{1}{2} \rho_2 v_2^3 A_2 = \frac{1}{2} \rho_1 v_1^3 A_1 \left(1 - \frac{v_2^2}{v_1^2}\right). \tag{7.72}$$

Because $\dot{W}_m = F_m v$ must hold, comparison of Eqs. (7.71) and (7.72) implies $v = (v_1 + v_2)/2$. Replacing $\rho_1 v_1 A_1$ by $\rho v A$ in Eq. (7.72) and then $v$ by the average of $v_1$ and $v_2$, gives

$$\dot{W}_m(v_2) = \frac{1}{4} \rho A (v_1^2 - v_2^2)(v_1 + v_2) - \frac{b}{2^{n+1}} (v_1 + v_2)^{n+1}, \tag{7.73}$$

In applications, $v_1$ is given by the incoming wind, while $v_2$ can be controlled by the rotor design and blade angle adjustment (*pitch-control*). This is why $\dot{W}_m$ is written here as a function of the control parameter $v_2$. It is common to introduce the velocity ratio $\lambda = v_2/v_1$ and define the efficiency as a function of $\lambda$

$$\eta(\lambda) = \frac{\dot{W}_m}{\dot{W}_1} = \frac{1}{2}(1 - \lambda)(1 + \lambda)^2 - \kappa(1 + \lambda)^{n+1}, \tag{7.74}$$

with the abbreviation $\kappa = b v_1^{n-2}/\rho A 2^n$. A graph of $\eta$ as a function of $\lambda$ is shown in Figure 7.12 (b) with Newtonian friction, $n = 2$, and different values of the loss coefficient $\kappa$. Consider first the ideal wind turbine, $\kappa = 0$. Stopping the wind to $v_2 = \lambda = 0$ has a 50% efficiency. Where is the energy going? Note that even if $v_2$ is virtually vanishing, the mass current is not stopped: the mass continuity Eq. (7.70) is still valid with a finite value of $\rho_2 v_2 A_2$. In the limit $v_2 \to 0$, one must have $A_2 \to \infty$ (for $\rho_2 \approx$ constant). Furthermore, $v = v_1/2$ and thus $A_1 = A/2$. The harvested power, according to Eq. (7.72), is the total power coming through the upstream area $A_1$, which is only half of the whole wheel area $A$. If $v_2$ is increased from zero, in Eq. (7.72) there is a reduction factor $(1 - \lambda^2)$ which is quadratic in $v_2$, but the prefactor $\rho_1 A_1$ increases linearly with $v_2$ at constant $A$. There is thus a maximum value of $\eta$. Solving $d\eta/d\lambda = 0$ for $\lambda$ gives $\lambda = 1/3$, or

$$v_2 = \frac{1}{3} v_1. \tag{7.75}$$

The maximum efficiency is

$$\eta^{(max)} = \frac{16}{27} \approx 60\% \tag{7.76}$$

The two last equations constitute the famous *Betz law*. In the literature our $\eta$ is usually called *power coefficient* or *power factor*. Typical maximum values for real wind turbines depend on the rotor technology and range from 30-50% for modern wind turbines.

Betz's law shows that even in the best case the total kinetic energy cannot be harvested by a wind wheel, since some energy will always escape, because mass conservation leads to a decrease of the effective area, $A_1$. If we take losses by viscous drag and turbulent eddies created by the blades into account, the efficiency is further reduced. This is illustrated in Figure 7.12 (b) by the curves for $\kappa > 0$. The optimum velocity $v_2$ is decreased from $v_1/3$, which must be expected because the loss increases with velocity in our simple model. The $v_2$-value where $\eta$ vanishes is decreased by the loss, which is also understandable. Note that in fact the derivation of Betz's law is not based on the assumption of lift force, so it is general. However, it turns out that in practice the efficiencies for drag force based turbines are smaller than the Betz limit.

## 7.5.2  DRAG-FORCE BASED FREE FLUID-FLOW POWER

The hydrological cycle of water circulation is part of the geo-solar-thermal energy-conversion system, and you know that it is one of the oldest sources for mankind's energy harvesting. Let's have a look at a flow with velocity $v_1$, which drives a water wheel with a horizontal axis above the water level on a river, as shown in Figure 7.11 (c). If we assume a Newton friction law for the vane of area $A$, which rotates with velocity $v$, the drag force becomes

$$F_d = \frac{c_d}{2}\rho A(v_1 - v)^2 \tag{7.77}$$

where the friction coefficient $c_d$ of the dragged object is a function of its geometry. Typical values, which can be found in appropriate engineering tables, are between 0.1 and 2.5. Note that in Eq. (7.77), we had to take the relative velocity, $v_1 - v$. For convenience, we will introduce the ratio $\lambda = v/v_1$ again. Prior to writing down the expression for power, $F_d v$, it should be mentioned that the dragged object has to return, which requires some power to pull it back against a force $\tilde{F}_d$, which is of the same form as Eq. (7.77). Of course, this loss can be minimized, either by making the associated friction coefficient $\tilde{c}_d$ small (by using an appropriate shape, such as for the anemometer in Figure 7.11 (b)) or by making $v_1$ and/or $\rho$ small on the way back, such as for our water wheel, where the way back goes through the air as shown in Figure 7.11 (c). For the ideal case we can neglect the force $\tilde{F}_d$ and then obtain for the power

$$\dot{W}_m = F_d v = \frac{c_d}{2}\rho A v_1^3 (1 - \lambda)^2 \lambda. \tag{7.78}$$

We define the drag efficiency coefficient by

$$\eta_d(\lambda) = \frac{\dot{W}_m}{c_w \rho A v_1^3/2} = (1 - \lambda)^2 \lambda. \tag{7.79}$$

Besides an optimization of $c_d$, the power can be maximized by tuning the wheel velocity $v$ such that $\lambda = 1/3$. This is the same value as in Betz's law. However, here $v$ does not describe the downstream velocity of the fluid, but the velocity of the vane. If the radius $r$ of the wheel is constant, it means that the angular frequency $\omega = v/r$ must be controlled for power maximization. In engineering it is usual to introduce as a parameter the so-called *tip-speed ratio*, which is here nothing more than the value of $\lambda = \omega r / v_1$. Furthermore, while we found Eq. (7.76) for the wind wheel, we obtain now one fourth,

$$\eta_d^{(max)} = \frac{4}{27} \approx 15\%. \tag{7.80}$$

Even if we include the value of $c_d$ for comparison, the drag-force based device is less efficient than lift-force based devices, since in practice $c_d$ is smaller than 4 (for the upper and lower half-spheres in Figure 7.11 (b), for example, the $c_d$ values are roughly $0.3 - 0.4$ and $1.2 - 1.3$, respectively [Qua08]). This is one of the reasons for the relative prevalence of the lift-force based technologies in wind-power applications.

### 7.5.3  IMPULSE BASED DUCTED FLUID-FLOW POWER

If the flow is ducted such that it cannot escape and bypass the turbine, a Betz-type of efficiency limitation does not appear in the usual turbines for hydro-power, gas, and steam power plants. Of course, the power conversion part in reactive turbines associated with heat will always be limited from above by the Carnot limit. Let us consider here a simple impulse turbine prototype, namely, the Pelton turbine illustrated in Figure 7.13. Assume a jet with velocity $v_{fluid} = v_1$ and incompressible density $\rho$, impinging the vane with velocity $v < v_1$, and being elastically reflected back with velocity $v_2$. Mass conservation implies that the impinging fluid mass in the moving system of the vane equals the back-scattered mass, i.e., $v_1 - v = v - v_2$. This yields $v_2 = 2v - v_1$. The force acting on the vane of area $A$ can again be obtained from $J_m(v_1 - v_2)$ (see the discussion before Eq. (7.71)), but note that here $A$ is given). One obtains

$$F = \rho A v_1 (v_1 - v_2) = 2\rho A v_1 (v_1 - v), \tag{7.81}$$

such that the power transferred to the turbine experiencing a torque $M = Fr$ and rotating with $\omega = v/r$ becomes

$$P = Fv = M\omega = 2\rho A v_1 (v_1 - v) v. \tag{7.82}$$

The efficiency with this idealized model becomes thus (see Section 7.5.1)

$$\eta = \frac{P}{\rho A v_1^3 / 2} = 4\lambda(1 - \lambda), \tag{7.83}$$

with $\lambda = v/v_1 = \omega r / v_1$. The efficiency maximum is unity and occurs at $v = v_1/2$, such that the rejected fluid has zero kinetic energy in the laboratory frame ($v_2 = 0$) and just falls down. Idle rotation corresponds to $v = v_1 = v_2$, where the jet and the vane have the same speed, and no power transfer can occur.

## 7.5.4 REACTION BASED DUCTED FLUID-FLOW POWER

Real reaction (hydro-) turbines combine reaction and impulse forces and are more complicated than impulse (Pelton) turbines. That is why we will postpone a brief remark on real ones to the next subsection. Here, we have first a quick look at a simplified prototype reaction turbine to understanding the concept: the *lawn sprinkler turbine* [CC13]. It is sufficient to consider a single arm sprinkler of radius $r$, with constant pipe cross-section area $A$, and rotating with angular frequency $\omega$ as illustrated in Figure 7.11 (e). A generalization to several arms is straightforward (you just must not forget that also the fluid power inlet is divided up into the several arms). For simplicity, suppose that the sprinkler sprinkles the fluid tangential to the circular path of rotation, i.e., the nozzle is perpendicular to the arm radius. If $v_{fluid} = v_1$ denotes the velocity of the fluid injected into the sprinkler arm in the axis center, and $v = \omega r$ is the nozzle speed in sprinkler rotation direction, the velocity of the ejected fluid in the laboratory frame is $v_2 = v_1 - v$. The mass flow rate is $J_m = \rho A v_1$, hence the force acting on the nozzle is $F = J_m v_2$. The torque becomes is $M = Fr$, and thus the power

$$P = M\omega = \rho A v_1 (v_1 - v)v. \tag{7.84}$$

With $\lambda = v/v_1 = \omega r/v_1$ the efficiency becomes

$$\eta = \frac{P}{\rho A v_1^3/2} = 2\lambda(1 - \lambda), \tag{7.85}$$

i.e., half of Eq. (7.83). The maximum efficiency amounts to 50% and occurs still at $v = v_1/2$. However, the meaning is different. While for the impulse turbine the fluid ended at rest for maximum power (which is necessary for total energy transfer), here the maximum refers to an equipartition of the input energy to the load and the ejected fluid. If the latter is at rest ($v_2 = 0$), there cannot be any torque and the sprinkler turns at no-load, idle rotation. For transmitting a finite force, the fluid mass needs to carry a reaction force. If the sprinkler is at rest, the force is $F = \rho A v_1^2$ as one expects.

## 7.5.5 EFFICIENCY-POWER RELATIONS OF HYDRO TURBINES

Now, let us go beyond such idealizations and have a quick look on more realistic cases by including empirical losses and other limitations. With the gravitational energy density $\rho g h$ of the fluid with density $\rho$, the total power of the flow coming through a pipe to the inlet of a turbine is

$$P_{tot} = \rho g h \dot{V}. \tag{7.86}$$

The volume flow, $\dot{V}$ of a turbine usually has a lower limit $\dot{V}_{min}$, below which no energy can be converted, and an upper limit $\dot{V}_{max}$ that can be swallowed by

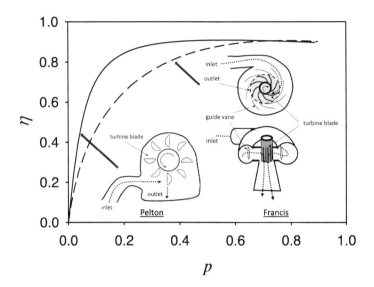

**Figure 7.13** Efficiency-power plot of different turbines: Pelton turbine (solid; $\lambda_m = 0.07$, $a_0 = 0.03$, $a_1 = 0.99$, $a_2 = 0.1$); Francis turbine (dashed; $\lambda_m = 0.095$, $a_0 = 0.18$, $a_1 = 0.63$, $a_2 = 0.31$); data from Ref. [Qua08].

the turbine outlet. It is convenient to introduce the normalized flow variable

$$\lambda = \frac{\dot{V} - \dot{V}_{min}}{\dot{V}_{max}}. \tag{7.87}$$

For $0 \leq \lambda \leq 1 - \lambda_m$, with $\lambda_m = V_{min}/V_{max}$, the turbine efficiency can often be phenomenologically approximated by an empirical fit [Qua08]

$$\eta = \frac{\lambda}{a_0 + a_1\lambda + a_2\lambda^2}. \tag{7.88}$$

The parameters $a_k$ ($k=0,1,2$) and $\lambda_m$ characterize the specific turbine. The mechanical power obtained from the turbine is

$$P = \eta P_{tot} = \eta\rho gh\dot{V} = \eta\rho gh\dot{V}_{max}(\lambda + \lambda_m). \tag{7.89}$$

By introducing the normalized power

$$p = \frac{P}{\rho gh\dot{V}_{max}} = \eta(\lambda)(\lambda + \lambda_m), \tag{7.90}$$

you can from Eqs. (7.88) and (7.90) plot the efficiency-power curves parameterized with $\lambda$. The result is shown in Figure 7.13 for two typical cases, a Pelton turbine and a Francis turbine with data from Ref. [Qua08]. The Francis turbine (shown in the inset of Figure 7.13) is an example for a typical realistic reaction turbine with a complex flow. It can only be understood in 3 dimensions, since the flow is whirling into the plane; it works with a combination of reaction and impulse forces. The figure shows the radial inflow in the top layer (white), the flow through the turbine (light gray) perpendicular to the plane, and the outflow (dark gray) behind the plane. The efficiency-power plot shows that the Pelton turbine has the better efficiency in the partial load regime. Nevertheless, the Francis turbine is probably the most prevalent turbine for hydro-power plants, mainly because of its large application area in terms of absolute power and of water drop height.

# 8 Performance Optimization

You have seen that energy conversion devices can often be characterized by their efficiency-power relation, $F(\eta, P) = 0$, where the function $F$ is determined by the physics of the device. Similarly, energy storage devices are characterized by their energy-power relation $F(E, P) = 0$ (Ragone plot). For the sake of convenience, we will in this chapter mostly, but not always, deal with the *normalized* power, $p = P/P_{max}$, and the *normalized* Ragone energy $e = E/E_0$, or the efficiency $\eta$. In general, the curves $F(\eta, p)$ (or $F(e, p)$) lead to an efficiency-power *trade-off issue*, and it is not a priori clear which is the optimum point of performance on the $\eta - p$-curves (see Figure 5.2 (b)). We will try to answer this question, which can provide us with the optimum design or the optimum operation point. *Design* has in our context a general meaning, as it is associated with a certain device property (quantified by design parameters), like geometry, material properties, etc., which can more or less be freely selected by the engineer from a well-defined value range.

A problem is that the quantitative meaning of performance is not so well-defined. Reconsider the dashed curve in Figure 1.1 (b) for a general case. If energy efficiency is the only important criterion for you, you chose the power value at maximum efficiency, and if you do not care at all about energy efficiency, you might want to maximize the power. There may be other constraints and requirements, like small volume, weight, environmental sustainability, acoustic noise, costs, long lifetime, and others. In any case, there is a certain ambiguity in what you finally want to optimize, and what will finally be the preferable operation point, $(\eta_{opt}, p_{opt})$ on the efficiency-power curve. Therefore, one cannot claim a generally valid solution for how to optimize technical devices. You should again look, in Figure 1.1 (b), at the part of the dashed efficiency-power curve with negative slope, which describes a set of *Pareto optimal* solutions of the multi-objective optimization problem of two conflicting goals, namely, maximizing power and efficiency. A Pareto optimum is defined as a point where one cannot improve one objective without worsening the other. Multi-objective optimization theory goes, however, beyond our purpose; we will instead introduce below a simple economic figure as the main objective that should be optimized on the efficiency-power curves.

In the remaining sections, four different optimization aspects for energy conversion and storage systems will briefly be outlined. The first refers to the determination of the efficiency-power curve of composite systems, e.g., a hybrid system. Then, we discuss entropy generation minimization, ecological function maximization, and net present value (NPV) maximization. The NPV is believed to be most general and, if done correctly, should cover the two other ones, which are applicable in special limit cases.

## 8.1  EFFICIENCY-POWER RELATIONS OF COMPOSITE SYSTEMS

What is the optimum efficiency $\eta(P)$ of an energy conversion system consisting of $K$ parts with their own efficiency-power relations $\eta_k(P)$, $k = 1, ..., K$? Let us consider this for a hybrid system ($K = 2$) as, e.g., shown in Figure 8.1. Well-known examples are hybrid cars, battery-supercapacitor systems, or the biological aerobic-anaerobic respiration hybrid in cells. The general case can be treated by building up the whole system step by step. Suppose the two subsystems are in parallel (for systems in series, the efficiencies just multiply) and each of them provides the power $P_k$. The total energy per time fed in is $\dot{W}_{in} = P_1/\eta_1 + P_2/\eta_2$, which is a function of $P_1$ for a given total power $P = P_1 + P_2$ received by the load. The best efficiency is thus obtained from maximization of $P/\dot{W}_{in}$, i.e.,

$$\eta(P) = \max_{P_1} \left( \frac{P}{P_1/\eta_1(P_1) + (P - P_1)/\eta_2(P - P_1)} \right) \tag{8.1}$$

with respect to $P_1$. If the power values are as usual limited by maximum values $P_{max,1}$ and $P_{max,2}$, the power of the composite system has the limit $P_{max} = P_{max,1} + P_{max,2}$. Equation (8.1) can be calculated for simple cases analytically, and for the general case, a numerical calculation with a computer is quickly done, particularly if the power values are limited. Let us illustrate this for the simple example where

$$\eta_k(P) = \eta_k^{(0)} \quad \text{for} \quad 0 < P \leq P_{max,k} \tag{8.2}$$

and $\eta_k(P) = 0$ otherwise, as shown by the constant solid lines in Figure 8.1 (b). Assume that $\eta_1^{(0)} > \eta_2^{(0)}$ and $P_{max,1} < P_{max,2}$. It is clear that in this case, $\eta(P) = \eta_1$ for $0 < P < P_{max,1}$. For $P_{max,1} < P < P_{max,1} + P_{max,2}$, the subsystem 1 runs on maximum power $P_{max,1}$, while the rest, $P - P_{max,1}$, is provided by subsystem 2. This yields, in the second $P$-range,

$$\eta(P) = \frac{P}{P_{max,1}/\eta_1^{(0)} + (P - P_{max,1})/\eta_2^{(0)}}. \tag{8.3}$$

The resulting efficiency is plotted as a function of the normalized power in Figure 8.1 (b) as a solid curve labeled by $\eta$. For illustration of a more general case, the same is done for two different parabolic efficiency-power relations (e.g., Eq. (1.2)). The efficiencies of the two sub-systems and the numerically calculated optimum efficiency of the hybrid are shown by dashed curves analogous to the previous example. Again, $\eta(P) = \eta_1(P)$ in a finite $P$-interval, which is defined by the condition that $\eta_1(P) \geq \eta(P)$, with $\eta(P)$ from Eq. (8.1); the point is indicated by the triangle in the graph.

## 8.2  ENTROPY GENERATION RATE MINIMIZATION

You learned that in real processes, entropy is generated which is equivalent to unwanted loss of exergy. Therefore, whenever you will be asked to design

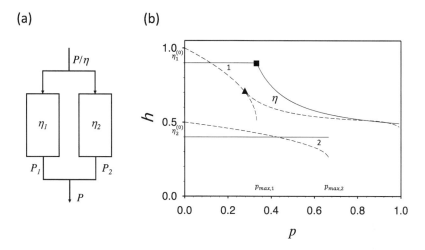

**Figure 8.1** Hybrid system: (a) parallel connection of two subsystems (1 and 2) with efficiency-power relations $\eta_k(P_k)$. (b) Maximum efficiency $\eta(p)$ of the hybrid system for two cases (with $p = P/(P_{max,1} + P_{max,2})$). Solid curves: constant subsystem efficiencies ($\eta_1 = 0.9$, $\eta_2 = 0.4$, see Eqs. (8.2) and (8.3)) with different maximum power values. Dashed curves: analogous illustration for parabolic efficiency-power relations of the subsystems. The symbols (square, triangle) indicate the onset of subsystem 2.

a technical device, an appliance, or an engine of whatever kind, you should check whether a design which minimizes the entropy generation rate, $\dot{S}$, is appropriate. While for given power constraints it is impossible to reduce $\dot{S}$ to zero (we skip the subscript *gen*, which was used in Section 3.1.2), but a reduction to a minimum is certainly better than doing nothing - at least if the cost of energy is an issue. The minimum can appear for a specific value of one or several design parameters due to a trade-off between different exergy destruction mechanisms. In the following, we shall discuss two simple examples for illustration. For more specific and realistic cases, see Refs. [Bej88, BTM96, Bej96].

## 8.2.1  ELECTRIC CURRENT DUCT

Suppose you have to select the cross section area $A$ of an electrical conductor with given length $L$, which conducts an electric current $I$ between a hot system at temperature $T_1$ and the ambient at $T_{amb}$. An illustration is shown in Figure 8.2 (a). Say, that the current is needed because you have to power an electric device inside the hot system, and you want to have the hot system thermally isolated from the environment. You realize this by applying a good thermal insulation everywhere, except for the connecting cable because it is made of

(a)                          (b)

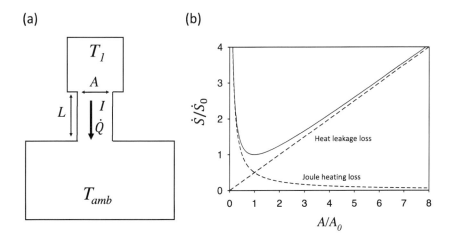

**Figure 8.2** Left: electrical/thermal link between two heat reservoirs. Right: individual (dashed) and total (solid) entropy generation rates of the current duct between the two reservoirs as a function of $A$ (normalized with respect to the minimum entropy generation values).

metal with a large heat conductivity. If you use a thin conductor, a great deal of Joule heat will be produced in the connection, and electrical power is lost, if it is too thin. On the other hand, if it is too thick, a fair amount of heat escapes from the hot system via the cable conductor. What is the optimum area $A$ if the electric conductivity $\sigma$ and the heat conductivity $\lambda_{th}$ of the conductor are fixed? The entropy generation rate due to Joule heating, $P = RI^2$, is given by Eq. (4.6)

$$\dot{S}_I = \frac{LI^2}{\sigma A T_{amb}} \quad (8.4)$$

where the electric resistance $R = L/A\sigma$ was used. Minimizing $\dot{S}_I$ requires, as one expects, a large area $A$. The entropy generation rate due to the heat loss is given by Eq. (4.8) with $\dot{Q}$ being the heat flux from the hot reservoir to the ambient,

$$\dot{S}_Q = \frac{A}{L}\lambda_{th}(T_1 - T_{amb})(T_{amb}^{-1} - T_1^{-1}), \quad (8.5)$$

which demands a *small* area $A$. Minimization of the total entropy generation, $\dot{S}(A) = \dot{S}_I + \dot{S}_Q$, as a function of $A$ gives with $d\dot{S}/dA = 0$ for the optimum area

$$A_0 = \frac{LI}{\sqrt{\lambda_{th}\sigma\Delta T\eta_0}}, \quad (8.6)$$

with $\Delta T = T_1 - T_{amb}$, and where $\eta_0 = 1 - T_{amb}/T_1$ is the Carnot efficiency associated with the two temperatures. The function $\dot{S}(A)$ is shown in Figure

8.2 (b). The entropy generation rate value $\dot{S}$ at minimum satisfies

$$T_{amb}\dot{S} = 2P = 2T_{amb}\dot{S}_Q. \tag{8.7}$$

It means that $P = \eta_0\dot{Q}$, i.e., the optimum corresponds to the balance of exergy destruction $P$, and loss of exergy by heat flow, $\eta_0\dot{Q}$. This result reminds us to always compare *exergies* when balancing power losses.

## 8.2.2  HEAT EXCHANGER

Heat exchangers are particularly suitable for applying the minimum entropy generation design principle [BTM96, Bej96, KL02]. Figure 8.3 (a) illustrates a heat exchanger system, where a cold fluid at temperature $T_2$ is flowing into a channel (or many channels, see Figure 8.3 (b)) that is in contact with a hot medium at $T_1$. Let $\dot{Q}_0$ be the power inflow of the cold fluid, which is subsequently loaded inside the heat exchanger with a heat per time, $\dot{Q}_1$, and flows eventually out with a heat power $\dot{Q}_2$ at elevated temperature $T_{1,c}$ ($T_2 < T_{1,c} < T_1$). This exit heat power can then be further used, e.g., in a reversible Carnot engine. Of course, a pump is needed which drives the fluid flow. Good heat transfer requires good contact of the fluid with the pipe wall, which has the unfortunate consequence of high friction and thus needs large pumping power $P$. Here we encounter the trade-off. Of course, $P$ is dissipated into heat, and in our model $\dot{Q}_2 = \dot{Q}_0 + \dot{Q}_1 + P$. Let us consider the outflow temperature as the reference temperature for the heat exchanger. Then, the entropy generation rate of heat exchange is $\dot{S}_Q = \dot{Q}_1(1/T_{1,c} - 1/T_1)$, and for the friction power it is $\dot{S}_f = P/T_{1,c}$. The total entropy generation rate is thus

$$\dot{S} = \frac{P}{T_{1,c}} + \dot{Q}_1\left(\frac{1}{T_{1,c}} - \frac{1}{T_1}\right). \tag{8.8}$$

This is the entropy generation rate that should be minimized. Let us quickly check whether this is indeed equivalent to minimizing the *loss of exergy* in the *complete* system, including the environment as a heat bath at ambient temperature $T_2$. Firstly, due to friction the pumping power $P$ ends up in heat at temperature $T_{1,c}$. Since the ambient temperature is $T_2$, the net exergy rate is $(1 - T_2/T_{1,c})P$. Hence exergy is lost at a rate $PT_2/T_{1,c}$. The pump power is not completely lost, but its exergy value is decreased. Secondly, the initial exergy of $\dot{Q}_1$ at $T_1$ is $\dot{Q}_1(1 - T_2/T_1)$, while what we really get out from the heat exchanger at $T_{1,c}$ is the exergy rate $\dot{Q}_1(1 - T_2/T_{1,c})$, which is of course less. The net exergy loss in the heat exchanger is thus $\dot{Q}_1T_2(1/T_{1,c} - 1/T_1)$. The total loss of exergy of both contributions is then

$$\frac{T_2P}{T_{1,c}} + \dot{Q}_1T_2\left(\frac{1}{T_{1,c}} - \frac{1}{T_1}\right) = T_2\dot{S}. \tag{8.9}$$

This should be minimized, and it is up to the constant factor $T_2$ equal to Eq. (8.8).

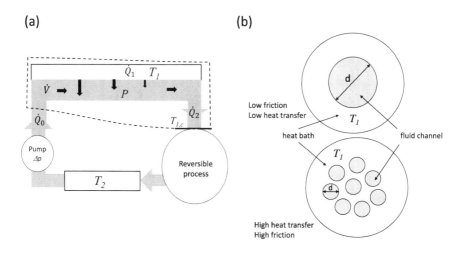

**Figure 8.3** (a) Heat exchanger model (see text). (b) Trade-off between friction losses (many narrow channels) and heat conduction losses (low heat transfer) for a heat exchanger with prescribed heat power delivery (constant total cross section (gray area); the heat conductivity of the heat bath (white area, $T_1$) is assumed to be very large).

Let us have a closer look at the typical design optimization trade-off problem, as indicated in Figure 8.3 (b). A good heat transfer associated with small $T_1 - T_{1,c}$ requires narrow flow channels with large surface-to-volume ratio. But this implies large flow friction, which enhances the pressure drop $\Delta p$ in the pump power term $P_p = -\dot{V}\Delta p$. A detailed analysis is rather complicated, because the spatial fluid-dynamics problem for heat and mass transfer must be solved. Even without working out the details, it should be clear that there is an optimum. Usually such flow problems are discussed in terms of non-dimensional parameters like *Reynolds, Prandtl, Nusselt numbers,* etc., which is useful because of universal scaling law. But we will skip this, for brevity, and allude here only to scaling with the channel diameter $d$ for the case depicted in Figure 8.3 (b). If mass flow rate (along the channel) and total flow cross-section have to be constant, the entropy production rate due to friction increases with decreasing $d$. On the other hand, the entropy production rate due to the heat transfer (perpendicular to the channel) at constant heat current $\dot{Q}_1$ is expected to increase with $d$, because it requires a larger temperature difference. If one can approximately describe these dependencies with power law exponents $k$ and $l$, the entropy production rate could have the form

$$\dot{S} = ad^{-k} + bd^l, \tag{8.10}$$

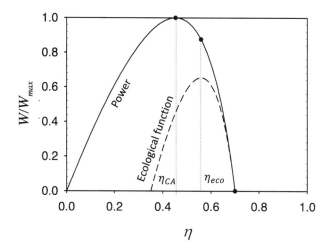

**Figure 8.4** Efficiency-power relation from Figure 5.2 (b), now with $\eta_{eco}$ associated with the maximum of the ecological function $\mathcal{E}(\eta)$ (cf. Eq. (8.13)).

with some constants $a$ and $b$. Minimization of $\dot{S}(d)$ is now straightforward, see Refs. [Bej88, BTM96, Bej96]) for details and many further applications.

It must be re-emphasized that the application of the minimum entropy generation design principle makes sense in cases where exergy is the relevant issue. The exergy loss is often only one out of several parts of the objective function.

## 8.3  THE ECOLOGICAL FUNCTION

Let us come back, for a moment, to endoreversible thermodynamics. Calculations of efficiencies of real power plants in the past have shown [CA75] that the values are often close to the Curzon−Ahlborn efficiency $\eta_{CA}$ (see. Eq. (5.19)), which refers to maximum power. The maximum efficiency, however, is still the Carnot efficiency $\eta_0$, which corresponds to minimum entropy generation. Independent of whether the true objective is economical, ecological, or of another nature, it is clear that the optimum efficiency will lie between the values $\eta_{CA}$ and $\eta_0$. The exact location depends on how the efficiency and the output power are weighted in the objective function. Suppose one is looking for a general trade-off between obtained and wasted exergies. Since $\dot{W}$ is the exergy rate obtained from the system and $T_2\dot{S}$ is the exergy destruction rate by irreversibilities, you may want to consider the function

$$\mathcal{E}_q(\eta) = \dot{W} - qT_2\dot{S}. \tag{8.11}$$

Maximization of this function refers to entropy generation minimization if $q$ is large, and power maximization if $q = 0$. Intermediate values of $q$ belong to other weightings of the two objectives. The $\mathcal{E}_{\mathrm{II}}$-maximization can be interpreted as *optimized utilization of exergy*. Appropriate values for $q$ might be justified with economics, like energy and investment costs. Nevertheless, we will consider here a simple case introduced by Angulo−Brown, the so-called *ecological function* [AB91],

$$\mathcal{E}(\eta) = \dot{W} - T_2\dot{S}, \qquad (8.12)$$

which corresponds to $q = 1$, and is shown in Figure 8.4. For the Novikov engine one obtains with Eqs. (5.16) and (5.24) $\mathcal{E} \propto (2\eta - \eta_0)(\eta_0 - \eta)/(1 - \eta)$. It has zeros at $\eta = \eta_0 = 1 - T_2/T_1$ and $\eta = \eta_0/2$, and a maximum in between. The latter is obtained from solving $d\mathcal{E}(\eta)/d\eta = 0$, which yields

$$\eta_{eco} = 1 - \sqrt{\frac{T_2}{T_1}}\sqrt{\frac{T_1 + T_2}{2T_1}}. \qquad (8.13)$$

The resulting efficiency lies obviously in between, the middle roughly, of the Curzon−Ahlborn and Carnot efficiencies, $\eta_{CA}$ and $\eta_0$, respectively.

## 8.4 ECONOMIC OPTIMIZATION

For a rational decision maker (*homo economicus*), the optimum of an *industrial* technical device is eventually defined via an economic measure that can characterize the *overall utility*. We emphasize here *industrial* devices - which actually means investment goods - because this is different for consumer goods, where irrationality and emotions may play a significant role in decisions. In economics most values are measured in terms of monetary quantities, and it is reasonable to have a brief look first at some basics of monetary values. You can find more details on economic optimization of energy devices in Refs. [BTM96, Ruf17, CO02].

### 8.4.1 NET PRESENT VALUE

The *net present value* (NPV) of an arbitrary object (e.g., an income, a technical device, a project, a property, etc.,) is the total monetary value of all cash flows associated with this object during its life-cycle, evaluated at one *specific* time. In order to understand the last clause of the previous sentence, one must be aware that equal nominal monetary values at different times are generally not equal in real values. Clearly, $x = 1\$$ now and $x = 1\$$ in one year do not necessarily have the same value. The reason is that lending or borrowing money is usually associated with a finite interest rate for the creditor. In our context, it is called the *cost of capital* for a company that borrows money for an investment, e.g., in order to develop a technology. We will not go into the details of how the cost of capital is determined and on which parameters it depends. Let us just assume a given yearly *technical interest rate r*. Then, 1\$ in one year is worth today only $1\$/(1+r)$. In order to make an economic decision today based on information on costs and benefits distributed in the future (that is a time series of cash flows), you first have to *discount* all monetary values to a fixed time, say to the value today, and then you sum them up. Because the future is never certain, probabilities come into play here and the so-called *utility functions* should be introduced, which describe the risk behavior of the decision maker. However, we will assume a risk-neutral decision maker, which is often reasonable to assume for industrial investment markets, and which allows us to skip the introduction of utility functions for energy conversion and storage devices (maybe this is not fully true for electrical cars which live in a rather irrational consumer market). Now, we can just use the monetary value as the objective to be maximized. If the value accumulating during year $n = 1, 2, \ldots$ is $x_n$ (the sign determines whether it is income or expenditure), the *present value* of all contributions is thus

$$PV(x) = \sum_{n=1}^{\infty} \frac{x_n}{(1+r)^{n-1}}. \tag{8.14}$$

For example, the $x_n$ might be yearly incomes from a power plant, and $PV(x)$ is then the present value of the total income. By definition, the value at $n = 0$ will be reserved for values *at* time zero, i.e., the initial *investment costs*. The value for $n = 1$ accumulates during the first year and is not discounted. Maybe that is all a bit simplistic, but acceptable for showing the concept.

The word *net* in the abbreviation NPV reflects the inclusion of all benefits and costs. In the framework of economic optimization of technical devices, one usually separates

$$NPV = B_{LC} - C_{LC}, \tag{8.15}$$

where $B_{LC}$ and $C_{LC}$ are the present values of the benefits and costs during the whole life-cycle of the device. We define the costs with a positive sign. They are further divided into *investment costs* and in repeatedly occurring *operation and maintenance costs*, $C_{LC} = C_{inv} + C_{op}$. The latter includes energy costs, labor costs, disposal costs, etc.. We just take them all in one.

Economic optimization of technical devices simply means

$$NPV = \text{maximum.} \tag{8.16}$$

It is clear that there is the additional constraint $NPV \geq 0$. Otherwise you would not invest since you loose money. In fact if you have different alternative opportunities, you should take the one with the largest NPV; sometimes one considers so-called *opportunity costs* associated with the virtual benefit of unrealized alternatives. The optimization criterion (8.16) is a main principle which is believed to explain a great deal of economical behavior. It is, however, fair to mention that other quantities are also considered. An example is the so-called *internal rate of return*, $r_{IRR}$. It can be understood by considering

$NPV(r)$ as a function of the interest rate $r$. For $r \to \infty$, the NPV is usually negative, as it is dominated by the investment costs, and the present value of all future net incomes vanishes. If $r$ decreases, the NPV (hopefully) crosses zero at a certain value of $r$, which is the internal rate of return: $NPV(r_{IRR}) = 0$. A necessary condition (analogous to $NPV > 0$) for a decision is thus that $r_{IRR}$ is larger than the true cost of capital, and the optimization strategy is to maximize the $r_{IRR}$. We will focus exclusively on the NPV.

## 8.4.2 POWER-EFFICIENCY TRADE-OFF

The aim of this section is to show with an example, how technical devices can be economically optimized by applying the $NPV$-maximization concept. For illustration, we consider a battery which is characterized by a Ragone plot or an efficiency-power relation [CO02]. Suppose that the investment costs depend on the size $N$ (with units, e.g., $m^3$, $kg$, or number of cells, etc.,) and the specific costs $c_N$ (units $\$/m^3$, $\$/kg$, or $\$/cell$, etc.,) in the form

$$C_{inv} = C_{inv}^0 + c_N N, \tag{8.17}$$

where $C_{inv}^0$ is a size-independent part. We also assume that the same amount of operation costs, $C_{op,n}$, accumulates in each year $n$ ($n = 1, 2, ..., \tau$) with constant interest rate $r$ during the lifetime $\tau$ of the energy storage device. The *present value* of the operation costs is

$$C_{op} = C_{op}^0 + c_e d \int_{\text{one year}} dt \, \frac{P_{req}}{\eta}, \tag{8.18}$$

where $C_{op}^0$ is energy independent, and the second part represents the cost of the primary energy needed to charge the energy storage device. The integral, which calculates the required energy for a given power demand $P_{req}(t)$ (units W), needs only to be done over one year, because it is supposed to be the same for every year. The amount of primary energy needed is of course larger than the secondary energy, namely, by a factor one over the total efficiency (*round trip efficiency*). This energy needs to be multiplied with the cost of energy $c_e$ (with units $\$/J$). Finally, the discounted sum over all years during the life-cycle leads, according to Eq. (8.14), to a geometrical series (because the $x_n$ are constant). Therefore a term $d = (\nu^\tau - 1)/(\nu - 1)$, with $\nu = 1/(1+r)$, appears in Eq. (8.18). Note that $d = d(\tau)$ is a function of the lifetime, which generally depends on the operation conditions. For example, if $p = P/P_{max} \to 1$, the lifetime might become short because large internal losses can lead to degradation of the device. For simplicity we will neglect this fact, although it is straightforward to take it into account if you want.

The net present benefits are similarly given by

$$B_{LC} = b_e d \int_{\text{one year}} dt \, P_{req}, \tag{8.19}$$

where the benefit (price) per output energy, $b_e$, appears (unit of $b_e$ is again $\$/J$). The net present value, $NPV = B_{LC} - C_{inv} - C_{op}$ has to be maximized under the constraint $NPV \geq 0$. Note that NPV-optimization means cost minimization only if $B_{LC}$ can be disregarded.

In the following, we determine for an example the optimum size $N$ of an energy storage device with the help of the Ragone energy-power relation $e(p)$, or the efficiency-power curve $\eta(p)$. We assume that $e$ is equal to the discharge efficiency. If they differ (remember Eq. (6.5)), you have to decide which one is relevant for your optimization purpose. The round trip efficiency can be written as $\eta = \eta_c e(p)$, where $\eta_c$ is the $p$-independent charging efficiency. You can now express the $NPV$ as a function of the design parameter $N$ by using $P = P_{req}/N$. $P$ is defined here as the *specific* power in units $W$ per cell, mass, or volume etc.. The required power demand, $P_{req}$, has units $W$. In dimensionless units one has $p = P_{req}/NP_{max}$, where $P_{max}$ is the maximum power of a unit, e.g., a cell. Hence one can use either $p$ or $N$ as the independent variable for optimization. This shows that determining the optimum number of cells is here the same as determining the optimum operation point on the $e(p)$-curve. Because $p \leq 1$, the minimum size is $N_{min} = P_{req}/P_{max}$. Optimization means $NPV =$ maximum. If we disregard terms independent of the design variable $p$ (or $N$), we get from a combination of Eqs. (8.15)−(8.19)

$$NPV = b_e d \int_{\text{one year}} dt \, P_{req} - c_e d \int_{\text{one year}} dt \, \frac{P_{req}}{\eta} - c_N N. \qquad (8.20)$$

Let us use $p$ as the variable and assume that $P_{req}$ is constant in time during the periods of discharge. The required energy per year is $\alpha P_{req}$ with utilization time $\alpha$ per year. During the period $1y(1 - \alpha)$, the energy storage device is charged with charging efficiency $\eta_c$, or stands still. In this simple case, the relevant, i.e., $p$-dependent part of the NPV (8.20) is given by the costs only, and can be written as

$$NPV(p) = \frac{c_N P_{req}}{P_{max}} \left( (\mu - \frac{1}{e})K - \frac{1}{p} \right) \qquad (8.21)$$

with dimensionless parameters defined by

$$K = \frac{c_e \alpha d P_{max}}{c_N \eta_c} \qquad (8.22)$$

and

$$\mu = \frac{b_e \eta_c}{c_e}. \qquad (8.23)$$

The parameter $K$ is, roughly speaking, a measure for the ratio of energy costs to investment costs. The parameter $\mu$ is a measure for the ratio of energy benefit (the energy sales price) to energy costs (taking into account the charging loss). If we assume that $d$ is $p$-independent (an assumption which you should scrutinize if you discuss a real case), maximum $NPV$ means

$$\frac{1}{Kp} + \frac{1}{e(p)} = \text{minimum !} \tag{8.24}$$

Besides the variable $p$, the only model parameter remaining in this optimization problem is $K$. If $e(p)$ is a decreasing function, a local minimum of Eq. (8.24) may be expected. Minimization leads to

$$Kp^2\frac{de}{dp} + e^2 = 0. \tag{8.25}$$

From Eq. (8.25) one finds the relation between $K$ and the optimum operation point $p$:

$$p_{opt} = \frac{1}{2}\left(2 + K^{-1} - \sqrt{4K^{-1} + K^{-2}}\right)\sqrt{4K^{-1} + K^{-2}}, \tag{8.26}$$

for the ideal battery without leakage, i.e., with $e = e_b$ from Eq. (6.16) for $R_L \to \infty$. For the sometimes found approximation $e = 1 - p$ for energy-power relations, Eq. (8.25) leads to the simpler expression

$$p_{opt} = \frac{1}{1 + \sqrt{K}}. \tag{8.27}$$

For every $K$ there is an optimum operation point on the Ragone curve or the efficiency-power curve. The relations $p_{opt}(K)$ and $e_{opt}(K) = \eta_{opt}/\eta_c$ are shown for the above examples in Figure 8.5. If investment costs dominate operation costs ($K \ll 1$), the optimum operation point is at maximum power ($p \to 1$), and the size $N$ should be as small as possible ($N \to N_{min}$). In the other limit $K \to \infty$, the efficiency has to be maximized. In the case of equality ($K = 1$), one finds $(p, e) = (0.85, 0.69)$ for the battery and $(0.5, 0.5)$ for $e = 1 - p$.

We did not say much yet about $\mu$ that appears in Eq. (8.23). Just minimizing costs cannot provide information on the value of the NPV. In order to make the NPV positive, you can now determine a minimum $\mu$-value, i.e., the minimum price for which you sell the energy with a benefit.

There are many possible extensions of the NPV-maximization procedure, including $p$-dependent lifetime, complex power demand profiles $P_{req}(t)$, etc.. An exercise that may be relevant in practice refers to optimizing the NPV for a *size dependent power* $P_{req} = P_{req}^{(0)}(1 + \kappa N)$, where $P_{req}^{(0)}$ and $\kappa$ are given constants. It may describe the case of battery-driven electric vehicles that must carry their battery with them. These remarks should be sufficient to achieve a first insight into the basic concept of economic optimization; real cases are always individual, and can be very complex and unwieldy.

(a)                                           (b)

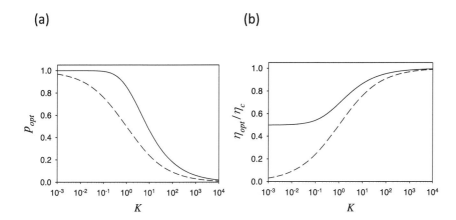

**Figure 8.5** Optimum normalized power values (a) and (discharge) efficiency values (b) as a function of $K$ (ratio of energy and investment costs) for energy storage devices with different efficiency-power relations. Solid: Eq. (8.26); dashed: Eq. (8.27).

# References

AB91.  F. Angulo-Brown. An ecological optimization criterion for finite-time heat engines. *Journal of Applied Physics*, 69(11):7465–7469, 1991.

Bej88.  A. Bejan. *Advanced Engineering Thermodynamics*. John Wiley & Sons, New York, USA, 1988.

Bej96.  A. Bejan. Entropy generation minimization: The new thermodynamics of finite-size devices and finite-time processes. *Journal of Applied Physics*, 79(3):1191–1218, 1996.

BSJB01.  T. Burton, D. Sharpe, N. Jenkins, and E. Bossanyi. *Wind Energy Handbook*. John Wiley & Sons, UK, 2001.

BTM96.  A. Bejan, G. Tsatsaronis, and M. Moran. *Thermal Design and Optimization*. John Wiley & Sons, New York, 1996.

CA75.  F. Curzon and B. Ahlborn. Efficiency of a carnot engine at maximum power output. *American Journal of Physics*, (43):22–24, 1975.

CC00.  T. Christen and M. Carlen. Theory of ragone plots. *J. Pow. Sources*, (91):210–216, 2000.

CC13.  Y. Cengel and J. Cimbala. *Fluid Mechanics Fundamentals and Applications: Third Edition*. McGraw-Hill Higher Education, NY, USA, 2013.

CCO99.  T. Christen, M. Carlen, and C. Ohler. Energy-power relations of supercaps from impedance spectroscopy data. In *9th Seminar on Double Layer Capacitors and Similar Energy Storage Devices, Dec. 99, Deerfield Beach*, 1999.

CF14.  T. Christen and Kassubek F. Entropy production moment closures and effective transport coefficients. *Journal of Physics D: Applied Physics*, 47(36):363001, 2014.

Cha57.  P. Chambadal. *Les centrales nucleaires*. Number 4. Armand Colin, 1957.

Che05.  G. Chen. *Nanoscale Energy Transport and Conversion: A Parallel Treatment of Electrons, Molecules, Phonons, and Photons*. MIT-Pappalardo Series in Mechanical Engineering. Oxford University Press, UK, 2005.

Chr06.  T. Christen. Application of the maximum entropy production principle to electrical systems. *Journal of Physics D: Applied Physics*, 39(20):4497, 2006.

CO02.  T. Christen and C. Ohler. Optimizing energy storage devices using ragone plots. *J. Pow. Sources*, (110):107–116, 2002.

Dat97.  S. Datta. *Electronic Transport in Mesoscopic Systems*. Cambridge Studies in Semiconductor Physics. Cambridge University Press, 1997.

DLNRL14.  R.C. Dewar, C.H. Lineweaver, R.K. Niven, and K. Regenauer-Lieb, editors. *Beyond the Second Law*. Springer, New York, 2014.

DV00.  A. De Vos. *Thermodynamics of Photovoltaics*, pages 49–71. Springer, New York, 2000.

DV08.       A. De Vos. *Thermodynamics of Solar Energy Conversion*. John Wiley & Sons, U.K., 2008.

HBS97.      K. Hoffmann, J. Burzler, and S. Schubert. Review article: endore-versible thermodynamics. *J. Non-Equilib. Thermodyn.*, (22):311–355, 1997.

HD13.       A. Hughes and B. Drury. *Electric Motors and Drives: Fundamentals, Types and Applications*. Elsevier Science, 2013.

Hil60.      T.L. Hill. *An Introduction to Statistical Thermodynamics*. Addison-Wesley Series in Chemistry. Dover Publications, 1960.

Hug10.      A. Huggins. *Energy Storage*. Springer, New York, 2010.

Ich94.      Masakazu Ichiyanagi. Variational principles of irreversible processes. *Physics Reports*, 243(3):125 – 182, 1994.

KK04.       S. Kulkarni and S. Khaparde. *Transformer Engineering, Design and Practice*. CRC Press, Taylor and Francis Group, 2004.

KKL$^+$12.  J. W. Kolar, F. Krismer, Y. Lobsiger, J. Muhlethaler, T. Nussbaumer, and J. Minibock. Extreme efficiency power electronics. In *2012 7th International Conference on Integrated Power Electronics Systems (CIPS)*, pages 1–22, March 2012.

KL02.       S. Kakac and H. Liu. *Heat Exchangers, Selection, Rating, and Thermal Design*. CRC Press, London, 2002.

Kre81.      H. Kreutzer. *Nonequilibrium thermodynamics and its statistical foun-dations*. Clarendon Press, Oxford, UK, 1981.

Kul15.      N. Kularatna. *Energy Storage Devices for Electronic Systems*. Elsevier, U.K., 2015.

LL13.       L.D. Landau and E.M. Lifshitz. *Statistical Physics*. Number 5. Elsevier Science, 2013.

LTAM80.     P.T. Landsberg, R.J. Tykodi, and Tremblay A.-M. Systematics of carnot cycles at positive and negative kevin temperatures. *J. Phys. A: Math. Gen.*, 13:1063–1074, 1980.

Mac09.      D.J.C. MacKay. *Sustainable Energy*. Without the Hot Air Series. UIT Cambridge, UK, 2009.

Mac15.      M.E. Mackay. *Solar Energy: An Introduction*. Oxford University Press, UK, 2015.

MR98.       I. Müller and T. Ruggeri, editors. *Rational extended thermodynamics*. Springer, New York, 1998.

MS06.       L.M. Martyushev and V.D. Seleznev. Maximum entropy production principle in physics, chemistry and biology. *Physics Reports*, 426(1):1 – 45, 2006.

Nov58.      I. Novikov. The efficiency of atomic power stations. *Journal Nuclear Energy II*, (7):125–128, 1958.

Qua08.      V. Quaschning. *Regenerative Energiesysteme (engl.: Understanding Renewable Energy Systems)*. Hanser, Munic, 2008.

Rag68.      D. Ragone. Review of battery systems for electrically powered vehicles. In *Mid-Year Meeting of the Society of Automotive Engineers*, May 20–24 1968.

Ruf17.      A. Rufer. *Energy Storage: Systems and Components*. CRC Press, Taylor and Francis, USA, 2017.

Smi17.      H. Smil. *Energy and Civilization, A History*. MIT Press, 2017.

Som96.      A. Sommerfeld. *Thermodynamics and Statistical Mechanics, Lectures on Theoretical Physics, Vol. V.* Academic Press, New York, 1996.

SQ61.       William Shockley and Hans J. Queisser. Detailed balance limit of efficiency of p-n junction solar cells. *Journal of Applied Physics*, 32(3):510–519, 1961.

Str14.      H. Struchtrup. *Thermodynamics and Energy Conversion.* Springer, Berlin, 2014.

The13.      A. Thess. Thermodynamic efficiency of pumped heat electricity storage. *Phys. Rev. Lett.*, (111):110602, 2013.

Wil61.      J. Wilks. *The third law of thermodynamics.* Oxford University Press, UK, 1961.

WW09.       P. Würfel and U. Würfel. *Physics of Solar Cells: From Basic Principles to Advanced Concepts.* Physics textbook. John Wiley & Sons, U.K., 2009.

# Index

Printed and bound by CPI Group (UK) Ltd, Croydon, CR0 4YY

17/10/2024

01775682-0020